Presenting
Your Findings

 Sixth Edition

Presenting Your Findings

A Practical Guide for Creating Tables

Adelheid A. M. Nicol and Penny M. Pexman

American Psychological Association • *Washington, DC*

Published by
American Psychological Association
750 First Street, NE
Washington, DC 20002
www.apa.org

To order
APA Order Department
P.O. Box 92984
Washington, DC 20090-2984
Tel: (800) 374-2721; Direct: (202) 336-5510
Fax: (202) 336-5502; TDD/TTY: (202) 336-6123
Online: www.apa.org/books/
E-mail: order@apa.org

In the U.K., Europe, Africa, and the Middle East, copies may be ordered from
American Psychological Association
3 Henrietta Street
Covent Garden, London
WC2E 8LU England

Typeset in Sabon, Futura, and Univers by Circle Graphics, Columbia, MD

Printer: United Book Press, Baltimore, MD
Cover Designer: Naylor Design, Washington, DC

The opinions and statements published are the responsibility of the authors, and such opinions and statements do not necessarily represent the policies of the American Psychological Association.

Library of Congress Cataloging-in-Publication Data

Nicol, Adelheid A. M.
 Presenting your findings : a practical guide for creating tables / Adelheid A. M. Nicol and Penny M. Pexman. — 6th ed.
     p. cm.
 Includes bibliographical references and index.
 ISBN-13: 978-1-4338-0705-3
 ISBN-10: 1-4338-0705-X
 ISBN-13: 978-1-4338-0706-0 (electronic version)
 ISBN-10: 1-4338-0706-8 (electronic version) 1. Statistics—Charts, diagrams, etc. I. Pexman, Penny M. II. American Psychological Association. III. Title.
 HA31.N53 2010
 001.4'226—dc22

                          2009038693

British Library Cataloguing-in-Publication Data
A CIP record is available from the British Library.

Printed in the United States of America
Sixth Edition

Contents

Preface

We decided to write the original version of this book when we were both graduate students at the University of Western Ontario. Adelheid was trying to summarize findings for a particular analysis. She clearly needed to create a table to do so, but she was struggling to figure out what the format of the table should be. She thought there must be a guidebook or some other source that would summarize table format and was surprised to find that no such book existed. Like other students, she had to search many periodicals and statistics textbooks before locating adequate models to guide her work. One evening, she was complaining over the telephone to a friend about how she was wasting too many hours looking for the proper way to make tables (when she could be doing other fun stuff like photocopying articles or running analyses). She mentioned that a reference guide definitely would be useful to have. Her friend (who was possibly tired of listening to her complain and was hoping that this comment would make her stop!) told her that she should write the book herself. Adelheid thought that this was a great idea. Unfortunately, her friend wanted no part in such a task. Frustrated, Adelheid went to tell Penny about her troubles (of course, Penny first had to listen to the entire drama that led to the idea). To Adelheid's delight, Penny was keenly interested in embarking on the project (even though we were both uncertain at the time what exactly we were getting into). We decided we could solve this problem. The happy ending to this story is the book you are reading, now in its revised edition.

In writing this book, we crafted the tool that we missed, and we hope it will be beneficial to all readers who, like us, have struggled to express statistical data correctly and elegantly. By presenting multiple examples of tables for the results of a wide range of statistical analyses, we have structured this book to make the reporting of findings easier.

Revised Edition

This revised edition of *Presenting Your Findings* was motivated by a number of developments, including the release of the sixth edition of the *Publication Manual of the American Psychological Association.*[1] The preparation of manuscripts for publication has changed dramatically since the original version of *Presenting Your Findings.* Most journals now require electronic submission of manuscripts and publish articles electronically. Software programs can easily be used to prepare tables and have relatively easy-to-use formatting functions. Some APA Style standards have changed. Furthermore, some reporting standards for statistics have changed. To reflect this changed environment, we have incorporated the following changes in the revised edition of this book:

1. Confidence intervals have been included, where appropriate.
2. Where possible (if it does not make the table too cluttered), exact probabilities are presented at two or three decimal places.
3. The font of all tables has been changed from Courier to the font style used in the sixth edition of the *Publication Manual.* (When submitting a manuscript, the preferred font is Times New Roman.)
4. All text that used to be underlined is now italicized.
5. Tables that are longer than a page do not need to include a "Table continues" note at the bottom of the page where the table ends and a "Table continued" note at the top of the page where the table continues in the manuscript version. This is because tables are now submitted electronically rather than in hard copy. When a table note does not fit on one page, however, include the "Table continues" and "Table continued" notes for clarity. (The typeset version will most likely include such notes, but they will be added to fit the style requirements of the publisher.) Only the stub head, column headings, and decked heads (if any) need to be repeated on subsequent pages.

Acknowledgments

In the creation of this book, many individuals were involved. We thank Tony Vernon, Robert Gardner, and David Stanley for their assistance in the development of the original version of the book. For the revised edition of the book, we thank all of the reviewers for their helpful comments and the staff of the American Psychological Association Books Department, especially Anne Woodworth Gasque, for helping us through its creation.

Adelheid Nicol dedicates this book to her husband, Yves Mayrand, and their three children, Ariane, Amélie, and Mathieu. Penny Pexman dedicates this book to her husband, Dave Pexman, and their two children, John and Kate.

---

[1]American Psychological Association. (2010). *Publication manual of the American Psychological Association* (6th ed.). Washington, DC: Author.

Presenting
Your Findings

Introduction

The table is an important medium for presenting the results of statistical analyses or for summarizing large amounts of textual information. Tables can be used to (a) summarize the important elements for the reader, (b) provide a quick and easy means of communicating information, (c) identify what is important (as presented and interpreted by the author), (d) present trends, and (e) provide other researchers with data that can be used for subsequent analyses. Tables can present large amounts of data and display a range too large to present well in a figure. When done well, tables are taken for granted as an effortless medium for studying their contents. When done poorly, they become a difficult puzzle for the reader to interpret.

This book provides the reader with numerous examples of tables that serve as models for what needs to be included for various types of statistical analyses. The examples are based on guidelines from the sixth edition of the *Publication Manual of the American Psychological Association*.[1] In this chapter, we first briefly describe how the tables in this book were created and then offer some suggestions for table design; a summary of APA Style issues that apply to tables; issues regarding the publication of tables on the Internet; a description of what this book is best used for (and what it is not intended to do); and finally, a description of how this book has been organized.

How We Created the Tables

As previously mentioned, we constructed the tables in this book to reflect the general guidelines for tables found in the sixth edition of the *Publication Manual* and refined those models by replicating the format of tables presented in published journal articles. For each type of statistical analysis, we looked at a broad sample of journal

---

[1]American Psychological Association. (2010). *Publication manual of the American Psychological Association* (6th ed.). Washington, DC: Author.

articles reporting that analysis. We sampled from journals reporting many areas of research, including animal learning, clinical psychology, cognition, developmental psychology, educational research, industrial/organizational psychology, neuroscience, nursing, psychiatry, and social psychology, with our sources of reference being APA's family of journals and other reputable journals. From these journals, we arrived at consensus about the common elements in tables for each type of statistical analysis. For some analyses, we present only one table format for results, whereas for other analyses, we present a range of formats that are commonly used. The tables included are those commonly found in the literature and for which clear and common standards in the literature exist.

Anatomy of a Table

Although many table examples are presented in the chapters that follow, a basic understanding of table components is important. If you find yourself grappling with how to present the results of a particularly complicated study, knowing the functional difference between the parts of a table, such as a spanner and a stub head, will help you in developing an acceptable modification. Table 1.1 illustrates the main parts of a table, which are defined as follows:

- *Table number.* Each table is given a table number. If more than one table is presented in a manuscript, they are numbered consecutively. Whole Arabic numbers (e.g., 1, 2, 3) are recommended for manuscripts that are prepared in APA Style. The table numbers presented here, such as 1.1, 2.3, and so on (the first number representing the chapter number and the second representing the order of the table within the chapter), is the style for printed APA books. For dissertations, reports, and book chapters, there may be different conventions to follow.
- *Table title.* The title is the description of the contents of the table. The title should not be a word-for-word repetition of column or row heads but rather should concisely express key groups and manipulations (see the box "What Makes a Good Table Title"). The title ideally summarizes what is being presented in the table (e.g., the group of variables of interest, the sample, and the statistical analyses such as correlation or analysis of covariance); thus, the reader scanning an article should be able to immediately identify what is being presented in the table, the variables of interest, and the type of analysis (if the table is to be used to present results).
- *Table spanner.* This is a subheading used for further division within the body of the table. A spanner is centered across the entire body of the table. Spanners are used to indicate variations in data that cannot be expressed in column heads or stub heads. Spanners should be used only if there is more than one category or grouping treated in the table. A single spanner is not necessary because the information applies to the entire table. Incorporate such information into the table title.

When to Use a Table

Tables allow complex data to be expressed in a tidy format. Putting research results in a table achieves two goals. First, details of the study are presented so they can be subjected to further analysis. Second, by removing long strings of data from the

■ Table 1.1.   Basic Components of a Table

table number ·····▶ Table X

table title ·····▶ *Numbers of Children With and Without Proof of Parental Citizenship*

column spanner: heading that identifies the entries in two or more columns in the body of the table

decked heads: heading that is stacked, often to avoid repetition of words in column headings

stub head: heading that identifies the entries in leftmost column

table spanner: heading that covers the entire width of the body of the table, allowing for further divisions

column heads: heading that identifies the entries in just one column in the body of the table

stub or stub column: leftmost column of the table; usually lists the major independent or predictor variables

cell: point of intersection between a row and a column

table body: rows of cells containing primary data of the table

table note: three types of notes can be placed below the table, which can eliminate repetition from the body of the table

	Girls		Boys	
Grade	With	Without	With	Without
		Wave 1		
3	280	240	281	232
4	297	251	290	264
5	301	260	306	221
Total	878	751	877	717
		Wave 2 ◀··· **table spanner**		
3	201	189	210	199
4	214	194	236	210
5	221	216	239	213
Total	636	599	685	622

Note. General notes to a table appear here, including definitions of abbreviations (see section 5.16 in the *Publication Manual*). From *Publication Manual of the American Psychological Association* (6th ed., p. 129), by the American Psychological Association, 2010, Washington, DC: Author. Copyright 2010 by the American Psychological Association.

[a]A specific note appears on a separate line below any general notes; subsequent specific notes are run in (see section 5.16).

*A probability note (*p* value) appears on a separate line below any specific notes; subsequent probability notes are run in (see *Publication Manual*, section 5.16, for more details on content).

What Makes a Good Table Title?

A good table title should be brief but clear and explanatory. It provides information about the results presented in the table without duplicating information presented in the table headings. A long title often looks awkward in relation to a small table and usually is the result of repeating information contained in the table. The basic content of the table should be easily inferred from the title.

Too general:

Relation Between College Majors and Performance

[It is unclear what data are presented in the table.]

Too detailed:

Mean Performance Scores on Test A, Test B, and Test C of Students With Psychology, Physics, English, and Engineering Majors

[This duplicates information in the headings of the table.]

Good title:

Mean Performance Scores of Students With Different College Majors[2]

text, one can approach the study from a broad perspective, using the body of the table to analyze trends and explore the implications of the results.

Tables should not be used when results can easily be expressed in text. If a table is unusually short (only a few columns or rows), it may be best to discuss the results within the text (this may not be the case for reports, theses, or dissertations). Likewise, tables should be limited to the expression of data that are directly relevant to the hypotheses in the research. Detailed results that are not directly relevant to the hypotheses may be included as a table in an appendix.

Table Design

Whether using a table in this book as a model or generating an original format for an analysis not covered here, one should remember that a good table should always be able to stand alone. That is, the reader should not have to refer to the text for basic information needed to understand the table. Simply looking at a table should provide enough information for the reader to grasp what is being presented.

Tables need some thought in design. A good table presents findings in a manner that makes it easy to read and easy to identify trends or anomalies. It is important to consider a simple layout—avoid complexity as much as possible. Some issues and questions to consider when designing a table are presented in the following checklist on table design.

---

[2]Adapted from *Publication Manual,* section 5.12, p. 133.

Table Design Checklist

Titles

☐ Is the title meaningful? If read out loud, does it make sense?

Column and Row Heads

☐ Are any column or row heads repeated unnecessarily?

☐ Is it clear to what column or row the various heads refer?

☐ Are row and column heads presented in a logical order? There are many ways of doing this; some include ordering them according to importance, ordering them according to their order of presentation as found in the body of the manuscript, ordering them from largest to smallest, or ordering them alphabetically. The order will depend on what is considered to be important, what the reader will naturally look for, and what makes for easy organization.

☐ Are long strings of words used in the heads when a few key words would suffice?

☐ Are columns approximately equally spaced to assist in distinguishing between them?

☐ Are certain columns grouped together (by spacing them closer together), when necessary, so that the reader looks at them together?

Table Spanner

☐ Would a table spanner assist in simplifying the table? Use spanners only if there is more than one category or grouping that requires a spanner. A single spanner is not necessary because the information applies to the entire table. Incorporate such information into the table title.

Number of Decimals

☐ Is it necessary to present data values to three to four decimal places, or will two or fewer suffice (to avoid clutter and to avoid presenting too much information)?

Abbreviations

☐ Are abbreviations and acronyms spelled out or explained and other basic information provided either in the table itself or in a table note?

Number of Tables

☐ Consider how many tables you should present—just one (with everything crammed into it) or a few (with the information distributed across them).

☐ When there are several tables in a manuscript presenting the results of one type of analysis (e.g., three tables of analyses of variance presented for three studies within a manuscript), the layout should be similar. This means that across tables, the column spanners, column heads, and stub heads should be similar and presented in the same order, whenever possible, and the data should be presented in a similar manner (e.g., same number of decimals, significance levels presented in the same way). Common presentation styles should be the same even across different types of tables within the same manuscript. For instance, notes, asterisks, and daggers should be used in a similar manner for all of the tables within the same manuscript.

☐ Tables should be referenced in the body of the manuscript and numbered in the order in which they appear.

Size of the Table

☐ Will the table be longer than one page? Very large tables, if thought not to be essential for understanding the body of the text, can be placed in an appendix. This is sometimes done to present the various studies used for meta-analyses, for instance, or to present the original or final items for a newly designed questionnaire. Extremely long tables, such as those presenting meta-analysis studies, are sometimes placed as supplemental material on the web so that they are easily accessible in electronic format.[3]

Table Placement

☐ Preferred table placement is immediately after discussion of that table within the text. APA recommends that authors submitting a manuscript to a journal place the tables after the references section, with each table beginning on a separate page. Many journals, theses, and reports require the tables to be integrated into the text rather than being placed at the end of the manuscript (with a note in the body of the text indicating where the table should be placed); this needs to be determined beforehand by reviewing journal or departmental guidelines. Note that even though the author may integrate a table into the manuscript, the typesetter will determine the most appropriate location for each table.

Description of Table

☐ When the table contents are described in the body of the manuscript, use the same terminology in the text as found in the table, and if highlighting certain elements (e.g., values) in the text, confirm that these values match those found in the tables.

☐ Always verify that the contents of the table are accurate to avoid confusion for the reader.

---

[3]See *Publication Manual*, section 2.10.

Although word-processing programs have table functions that permit the user to easily and rapidly create tables, once the table has been created, changing and reformatting can take a lot of time that could have been saved if a bit of planning and sketching had been done beforehand. A good table is easy to read and makes it easy for the reader to identify trends or anomalies. Creating such tables comes more naturally with experience and lots of practice. What can help when designing a table is to sketch out the table on paper to determine how the information is to be presented, examine similar tables (such as those found in this book), ask for opinions, and take a few days' break from the initial design so you can look at it with fresh eyes to determine if it is clear and effective.

APA Style Issues

Some APA Style issues that need to be considered when creating tables include the following (most are also mentioned in the Preface to this book).

Confidence intervals. Confidence intervals should be included for point estimates—for instance, means, correlations, or regression slopes.

Probability values. Exact probability values (e.g., $p = .03$) should be presented if space permits. These should be at two or three decimal places (values requiring more than three decimal places are usually presented as $p < .001$). If space does not allow the presentation of probability values in the body of the table, they can be presented in a probability note at the bottom of the table (see Table 1.1 for exact placement below the table). If probability notes are used, the symbol used to indicate a specific p value should remain the same across all of the tables in the manuscript and should be repeated in tables that present values significant at that level.

Font style. The font style (e.g., Times New Roman, Arial) and size (e.g., 12 points) of all tables should match the rest of the text. For APA publications, Times New Roman (12-point type size) is preferred. Check with the journal publisher or university for specific requirements.

Italics. The table title and statistical abbreviations should be italicized.

Line spacing. Tables can be typed using single or double spacing. Table titles can be typed using single-, one-and-a-half-, or double-spacing.

Tables that require more than one page. Tables that span more than one page should repeat the table stub head, column heads, and any decked heads on each page in which the table appears. If a note of a table does not fit on the page and it is the only thing that does not fit, put "Table continues" at the bottom of the page and "Table continued" at the top of the next page and include the note on the next page for clarity. If a table is in landscape format and it is too wide for the page, the table continues on the next page with the stub head and row heads and the new column heads for the remainder of the table.

Rules. Many text editing programs include a grid of lines (rules) when a table is created (i.e., each cell has rules or borders on all four sides). In APA publications, the use of gridlines is discouraged and authors are encouraged to limit the use of rules (see *Publication Manual,* section 5.17). Include rules at the top and bottom of the table (to separate the column heads from the results in the table), between decked heads, above table spanners, and as needed to distinguish totals for a column of values.

Stub head. The stub head should be centered in its column.

Column heads and spanners. All column heads and column spanners are centered over their columns.

Table spanners. Table spanners are centered over their columns. They are used only if there is more than one category or grouping that requires a spanner.

Alignment of material presented in the body of the table. Results are centered under their column head (this does not apply to text or words that are left justified in their respective columns or to numbers with decimals, which are aligned by decimal point). This centering is of less concern when a manuscript is being submitted for publication because the typesetter can center the numbers under the columns. However, for theses and lab reports, the student or researcher is responsible for doing this if the material must be presented in APA Style. Some programs have the ability to align numbers on the decimal automatically (along with the usual left, right, and center alignments), but some do not. If the program does not have this feature, you will have to add in spaces and align the decimals manually.

Abbreviations. Abbreviations for words or terms can be used in a table when they improve clarity and readability and can save space when a table has many columns or rows. In APA publications, the abbreviation must be defined in the title of a table (if appropriate), in the table itself, or in a note.

Empty cells. If nothing is to be put in a cell, leave it blank (do not put a dash, a bar, or X in the cell). If cells are empty because it was not possible to obtain data or the writer does not want to report the data in those cells, a long dash, or *em dash* (—), should be put in the cell accompanied by an explanation in a table note. Dashes used in the main diagonal of a correlation matrix (to represent the correlation of a variable with itself) do not need to be explained in a table note. Superscripts (specific notes) are used to explain data missing from specific cells (see Table 1.1 for placement of specific notes for a table).

It is important to note that APA Style is not universally required. The requirement to use APA Style will depend on the guidelines of the journal to which a manuscript is being submitted, the specifications of the instructor for a lab report for a course, or the specific formatting rules for a thesis or dissertation. The researcher will have to determine in each instance whether APA Style has to be followed.

Web Publishing

A great deal of the research literature is now read online rather than on paper. This has implications for how to present your tables. For instance, although individuals should not be discouraged from presenting a table in landscape format, it is much easier to read landscape format on paper than online (some readers may not know how to rotate the page online). Also, tables that extend over one page are more difficult to read online than on paper.

Some journals will accept additional tables that will be presented only online and not in a printed form. These kinds of supplemental materials can include tables that are exceptionally large or present additional information such as data sets (see the *Publication Manual,* section 2.13, for more on supplemental materials and online supplemental archives). It is up to the researcher to determine whether she or he wishes to create additional tables that will be viewed only as supplemental materials and whether the journal accepts such materials.

A Few Caveats

This book is not intended to be a statistics textbook. The goal was to provide a simple, visual presentation of tables along with straightforward research examples, not to provide recommendations on how to conduct analyses. The many complex issues relating to statistics are beyond the scope of this book and could not be adequately addressed here. Therefore, each chapter contains a brief description of the statistical analysis used in the sample tables but no advice regarding those statistical procedures (such as what to do about missing data or unequal Ns). These descriptions are intended to help the reader identify the analysis rather than to provide a thorough explanation of the statistical procedure. Many issues around statistics can be very complex, and we do not feel that these issues could be adequately addressed in this book. People who are using a particular statistic should be acquainted with the procedure or become acquainted by reading any one of numerous statistic books and references. Our book is meant to help those who understand the statistics they are using and want to know what is commonly presented—what information they should show. As an example, students running their first correlational study should understand what a correlation is, what the level of significance means, and so on. However, they may not know that one needs to present only one half of the correlation matrix in the results of a correlational analysis when presenting the variables intercorrelated with each other, and this book will provide this information. Similarly, this book will inform them that it is a good idea to present the means and standard deviations of the variables included in the correlational analysis. For more detailed explanations and descriptions of the various statistical procedures, we recommend reading the statistics textbooks referenced at the end of this chapter.

This book is also not about the art and design of communication through the use of tables. There are excellent books (including the *Publication Manual*) that address issues on the creation of tables and the various purposes of tables. If we addressed those issues our treatment would be redundant with many of those published books and also would detract from the primary purpose of this book—to specify the commonly identified important elements that should be presented when creating a table for presenting results. How that information should be best presented from a communications perspective (the "art" of table presentation) is not addressed in this book. We recommend the works by Bigwood and Spore[4] and Morgan, Reichert, and Harrison[5] for additional readings on these aspects of tables.

In some research, it can be beneficial to provide figures to illustrate results. However, few figures have been included in this book; figures are presented only in the chapters on cluster analysis (Chapter 13), discriminant function analysis (Chapter 15), and structural equation modeling (Chapter 20). The sister book to this one,

---

[4]Bigwood, S., & Spore, M. (2003). *Presenting numbers, tables, and charts.* New York, NY: Oxford University Press.
[5]Morgan, S. E., Reichert, T., & Harrison, T. R. (2002). *From numbers to words: Reporting statistical results for the social sciences.* Boston, MA: Allyn & Bacon.

Displaying Your Findings,[6] provides examples of presentation format for various figures, photographs, and conference posters.

The examples in this book were created for the purpose of illustrating how a table can be presented. The research study examples and the data for those examples are fictional; any resemblance to actual studies is purely coincidental. In addition, all of the measures identified in this book are fictional.

Note also that readers should not consider the design of the study examples in this book to be ideal. There may be better ways of investigating the research questions presented in the study examples. Statistics presented in the tables should not be considered the only way those data could have been analyzed; in many cases, the data could have been analyzed differently.

Finally, it is important to keep in mind that the tables in this book were created according to specifications in the sixth edition of the *Publication Manual.* Non-APA journals will have different specifications regarding the formatting of their tables, and individuals writing their theses or working on reports may have different typesetting rules to follow. Thus, the formatting of the tables may vary, but in many instances, the content of the tables in this book (e.g., what information should be presented for correlational analyses), and information on how to organize and structure a table provide a useful guide.

Organization of the Book

Each chapter in this book is devoted to a particular statistic (except Chapter 22, on word tables). The statistics selected were included because we believe them to be the most commonly used analyses. By selecting the statistics in this way, we hope to make the book useful for as many students and researchers as possible.

The chapters are presented by the name of the statistic they describe and in order of the complexity of the statistic. Word tables are presented at the end because they do not feature any statistics. Thus, many of the statistics commonly used by undergraduate and graduate students will be found at the beginning of this book. We hope that this will make it easy for the reader to find quickly a table model for whatever statistical analysis he or she is presenting. Some chapters are long because there are numerous ways to present the results or because several tables are required to present the results of that statistic. Some chapters are short because the results are presented with tables that are already illustrated in other chapters. If you are using this book to understand statistical presentation in tables, we recommend that you begin by looking at the chapters on presenting means (Chapter 3) and on frequency and demographics (Chapter 2) because the tables presenting these analyses form the basis for many other presentations.

To make this book easy to use, we have organized the chapters in a consistent format. There are five parts to each chapter: (a) a description of the statistical analysis, (b) an overview of the types of tables that are frequently presented for that particular analysis, (c) the "Play It Safe" table or tables identifying the most comprehensive presentation of the analysis, (d) one or more example studies that provide the context for the sample tables, and (e) the sample tables. The following paragraphs describe each of the five parts.

---

[6]Nicol, A. A. M., & Pexman, P. M. (2010). *Displaying your findings: A practical guide for creating figures, posters, and presentations* (6th ed.). Washington, DC: American Psychological Association.

What Is It?

A brief (one- to three-sentence) description of the particular analysis is provided to assist the reader to put the chapter in context. This description is not meant to provide a detailed and complex explanation of statistics. As mentioned earlier, for detailed explanations of the various statistics, we recommend a number of statistics textbooks at the end of this chapter.

What Tables Are Used?

One or two paragraphs describe the most commonly used table (or tables) for the results of the particular analysis.

"Play It Safe" Table

In many cases there are several alternatives presented for the format of the tables. Thus, in each chapter a "safe" format is indicated. This "Play It Safe" choice is comprehensive and thus would be appropriate if the writer wanted to be as thorough as possible and was not concerned with brevity.

Example

Most chapters contain at least one study example. In some chapters, several examples are presented. These examples are intended to be straightforward and therefore do not include especially complex variables or research methodology. The intent was to make the tables easy to understand by describing an example study and using fictional data. A researcher may need to extrapolate from the sample tables if his or her statistics are more complicated.

Accompanying the description of the example study is a list clearly identifying the independent and dependent variables. This is to help the reader understand and recognize the key elements of the study.

Sample Tables

After the description of the fictional example, one or more tables present the results of the statistical analysis. Each table adheres to the guidelines outlined in the sixth edition of the *Publication Manual*. Text notes, enclosed in shaded boxes, are located within, next to, or below some sample tables. These text notes are included to point out simple modifications that could be made to a table or to identify important aspects of a particular table. In many cases, alternate versions of the same table are presented to provide different examples of acceptable tables.

Conclusion

We hope this book simplifies the task of creating tables for research results. Our goal is to help researchers spend more time doing research and generating ideas and less time mulling over what should be included in their tables.

Additional Resources: Statistics

Grimm, G. L., & Yarnold, P. R. (Eds.). (1995). *Reading and understanding multivariate statistics*. Washington, DC: American Psychological Association.

Grimm, G. L., & Yarnold, P. R. (Eds.). (2000). *Reading and understanding more multivariate statistics*. Washington, DC: American Psychological Association.

Howell, D. C. (2010). *Statistical methods for psychology* (7th ed.). Belmont, CA: Wadsworth.

Tabachnick, B. G., & Fidell, L. S. (2007). *Using multivariate statistics* (5th ed.). Boston, MA: Pearson/Allyn & Bacon.

2

Frequency and Demographic Data

What Is It?

Frequency data are used to summarize the number of cases or instances of a particular characteristic or variable. Demographic information provides a summary of participant characteristics (e.g., age, occupation). (See section 2.06 of the sixth edition of the *Publication Manual of the American Psychological Association* for more on reporting research participant characteristics.)

What Tables Are Used?

Frequency data are included in a table only if they are particularly important in a study—for example, if the dependent measure is frequencies of particular behaviors. Demographic information usually is presented in a table if the participants are a special population (e.g., a clinical sample, an animal sample, or an organizational sample). Often, the data that are presented in a demographic table are frequencies (e.g., number of participants in each age group). It should be noted that demographic information can be displayed in either a frequency table or a table of means and standard deviations or in a combination of the two.

"Play It Safe" Table

For demographic data, the "safe" choice is Table 2.1 because it is more comprehensive. For a frequency table, the "safe" choice is Table 2.3.

Example

A group of researchers were interested in maternity leave in major corporations. Specifically, the researchers wanted to determine women's attitudes toward their corporations' maternity leave policies. The researchers surveyed 1,022 women who had taken maternity leave from corporations in major North American cities. The survey used asked the women questions about many aspects of their maternity leaves and corporation policies, including length, number of leaves per employee, and benefits during leaves. In summarizing their research, the researchers created a demographics table of the characteristics of the women surveyed (Table 2.1 or 2.2) and a frequency table to summarize their results (Table 2.3).

Variables for Example 2.1

1. Maternity leave policy
2. Attitude toward maternity leave policy

Table 2.1.

This is the "Play It Safe" table for demographic information.

Table X

Demographic Characteristics of Participants (N = 1,022)

Characteristic	n	%
Age at time of survey (years)		
20–29	244	24
30–39	534	52
40–49	132	13
50–59	112	11
Age at time of maternity leave (years)		
20–29	122	12
30–39	834	82
40–49	66	6
Highest education level completed		
High school	245	24
Undergraduate school	441	43
Graduate school	133	13
Professional school	203	20

■ **Table 2.1.** (continued)

Table X

Demographic Characteristics of Participants (N = 1,022)

Characteristic	n	%
Annual income ($)		
0–14,999	129	13
15,000–29,999	201	20
30,000–44,999	309	30
45,000–59,999	211	21
60,000–74,999	109	11
75,000–89,999	42	4
90,000–104,999	19	2
105,000+	2	<1
Length of leave (weeks)		
0–3	110	11
4–6	243	24
7–9	286	28
10–12	198	19
13–15	155	15
16–18	24	2
19–21	4	<1
22+	2	<1
Average pay received during leave (%)		
0–24	134	13
25–49	300	29
50–74	278	27
75–99	234	23
100+	76	7
Number of leaves taken		
1	502	49
2	322	32
3	159	16
4	39	4

Note. Totals of percentages are not 100 for every characteristic because of rounding.

Table 2.2.

Table X

Participant Characteristics (N = 1,022)

Characteristic	M	SD
Age at time of survey (years)	37.12	8.21
Age at time of maternity leave (years)	32.33	4.13
Years of education	15.34	3.04
Annual income ($)	38,723	15,201
Length of leave (weeks)	7.65	3.95
Average pay received during leave (%)	43	20
Number of leaves taken	1.66	0.71

Table 2.3.

This is the "Play It Safe" table for frequency information.

This table presents the frequency of responses for one of the survey questions.

Table X

Responses to Survey Question "What Would You Change About Your Employer's Maternity Leave Policy? (Choose One)"

Response	n	%
Nothing	33	3
Length of paid leave	145	14
Hiring of replacement for leave	22	2
Salary during leave	567	55
Stigma associated with leave	99	10
Benefits during leave	114	11
Layoff protection during leave	42	4

Note. N = 1,022. Total of percentages is not 100 because of rounding.

3

Means

What Is it?

A *mean* is a measure of central tendency. It is also called *the average*. For the purposes of this chapter, we define mean as the sum of a set of values divided by the number of those values. The standard deviation is frequently presented with the mean. The *standard deviation* is an indicator of the spread (variability) of the data. Both are often considered important for descriptive statistics (see Chapter 2 of this volume). Other statistics frequently presented with the mean include the *standard error of the mean* (the standard deviation of the sampling distribution of the mean) and the *confidence interval* (which provides an indication, described by upper and lower confidence limits that one can state with a certain percentage of confidence, e.g., such as 95% or 99%, that the population mean falls within the limits specified).

What Tables Are Used?

There are numerous ways in which means can be presented. Sometimes, in addition to the standard deviations (see Tables 3.1–3.3), the number of participants is included (see Tables 3.4 and 3.5). If there is little space, such as when there are many means, they may be presented without their standard deviations. It is important to consider that means often are presented in graphs rather than tables (note that no graphical examples are included in this chapter) or in tables in combination with other results such as coefficient alphas, intercorrelations, percentage of spoiled trials removed from analyses, and so on. Example 3.1 presents the results of the effects of stress on a single dependent variable, Example 3.2 presents the results of the effects of stress on two dependent variables and displays the inclusion of standard errors of means, and Example 3.3 presents the results of a different study and includes confidence intervals.

"Play It Safe" Table

The most comprehensive table includes the means, standard deviations (whether standard deviations or standard error of means are presented depends on the conventions of a particular discipline), and number of participants. If the number of participants in each cell is the same, it is not necessary to include this information (see Tables 3.1–3.3). If the number of participants varies from cell to cell, then this information would be useful (see Tables 3.4 and 3.5).

Example 3.1

Three species of male rats (Species X, Y, and Z), all of the same weight and age, were randomly assigned to one of two conditions. The two conditions represent two different stress environments. *Stress* is operationally defined here as exposure to 85-dB heavy metal music. Rats in the stressful environment were exposed to the heavy metal music 6 hr per day for 30 days. The rats in the control group were maintained in their usual environment (i.e., no exposure to any kind of music). The goals of this study were to examine the effects of a stressful environment on the amount of food eaten and to determine whether different species of rats are differentially affected by the stressful situation examined. The amount of food eaten by the different species of rats was recorded for 30 days.

There are two independent variables: species and stress environment. The independent variable species has three levels (Species X, Species Y, and Species Z). The independent variable stress environment has two levels (high stress [85-dB heavy metal music] and low stress [no music]). The dependent variable is the amount of rat chow eaten.

Variables for Example 3.1

Independent Variables

1. Rat species (X, Y, or Z)
2. Stress environment (high or low stress)

Dependent Variable

1. Amount of rat chow eaten (in grams)

Researchers often will present descriptive statistics before presenting their analyses. Tables 3.1 through 3.5 provide various alternatives for presenting means and standard deviations. In particular, Tables 3.4 and 3.5 include the numbers of participants.

Table 3.1.

The abbreviations used in the table can be changed from those shown if the data to be presented are mean reaction times, percentage of error rates, standard errors, probabilities of correct answers, percentages correct or incorrect, and so on.

Table X

*Mean Amount of Rat Chow Eaten (in Grams)
and Standard Deviations for Three Species
of Rats and Two Stress Conditions*

	High stress		Low stress	
Species	*M*	*SD*	*M*	*SD*
X	702.68	9.21	713.54	10.62
Y	721.39	14.76	724.76	15.98
Z	717.25	16.19	729.14	14.76

If there are a large number of means to present and space is a concern, the researcher could opt to omit the standard deviations.

This is a "Play It Safe" table of means and standard deviations when the number of participants in each cell does not vary.

Other information can be included in a table of means and standard deviations. For example, range, 95% confidence intervals, or coefficient alphas could be added in additional columns or rows.

Table 3.2.

Table X

*Condition Means in Grams for Species X, Y, and Z in
the Two Stress Conditions*

Species	High stress	Low stress
X	702.68 ± 9.21	713.54 ± 10.62
Y	721.39 ± 14.76	724.76 ± 15.98
Z	717.25 ± 16.19	729.14 ± 14.76

Note. Values are *M* ± *SD*.

This is a "Play It Safe" table of means and standard deviations when the number of participants in each cell is equal.

Table 3.3.

This is a "Play It Safe" table of means and standard deviations when the number of participants in each cell is equal.

Table X

Mean Amount of Rat Chow Eaten (in Grams) and Standard Deviations for Three Species of Rats Under High- or Low-Stress Conditions

Species	High stress	Low stress
X		
M	702.68	713.54
SD	9.21	10.62
Y		
M	721.39	724.76
SD	14.76	15.98
Z		
M	717.25	729.14
SD	16.19	14.76

Decimals are aligned under each column head. This centering is less of a concern when a manuscript is being submitted for publication because the typesetter can center the numbers under the columns.

■ Table 3.4.

Table X

Mean Amount of Rat Chow Eaten (in Grams) and Standard Deviations for Three Species of Rats Under High- or Low-Stress Conditions

Species	High stress	Low stress
X		
M	702.68	713.54
SD	9.21	10.62
n	45	49
Y		
M	721.39	724.76
SD	14.76	15.98
n	47	54
Z		
M	717.25	729.14
SD	16.19	14.76
n	58	47

■ Table 3.5.

Table X

Mean Amount of Rat Chow Eaten (in Grams) by Three Species of Rats in Two Stress Conditions

Species	High stress			Low stress		
	M	*SD*	*n*	*M*	*SD*	*n*
X	702.68	9.21	45	713.54	10.62	49
Y	721.39	14.76	47	724.76	15.98	54
Z	717.25	16.19	58	729.14	14.76	47

Example 3.2

This is the same study as described in Example 3.1, but here the effects of stress and species are examined on two dependent variables: (a) amount of rat chow eaten and (b) amount of pacing. Results are displayed in Table 3.6.

Variables for Example 3.2

Independent variables

1. Rat species (X, Y, or Z)
2. Stress environment (high or low stress)

Dependent variables

1. Amount of rat chow eaten (in grams)
2. Amount of pacing (in centimeters per minute)

■ **Table 3.6.**

Table X

Amount of Rat Chow Eaten (in Grams) and Amount of Pacing (Centimeters per Minute) by Three Species of Rats in Two Stress Conditions

| Species | High stress | | Low stress | |
	M	*SEM*	*M*	*SEM*
	Amount of rat chow eaten			
X	702.68	1.19	713.54	1.37
Y	721.39	1.90	724.76	2.06
Z	717.25	2.09	729.14	1.90
	Amount of pacing			
X	100.25[a]	2.80	65.23	1.10
Y	135.21	3.63	87.34	1.29
Z	98.87	2.20	45.32	1.22

Note. n = 60 for each cell. *SEM* = standard error of measurement.
[a]Number of rats in this cell = 58.

Here, instead of *SD*s, *SEM*s are presented. Whether *SD*s or *SEM*s are presented depends on the conventions of a particular discipline.

Example 3.3

Here a researcher wished to study various characteristics of nurses and midwives to determine if any of these characteristics can help distinguish the two groups. Various scales were included in this study to measure the following: satisfaction with work; commitment to work; organizational citizenship behaviors; attitude toward doctors; attitude toward nurses; attitude toward midwives; attitude toward alternative medicine; and nurturing personality trait, dominant personality trait, and controlling personality trait. Results are shown in Tables 3.7–3.9.

Variables for Example 3.3

 1. Satisfaction with work
 2. Commitment to work
 3. Organizational citizenship behaviors
 4. Attitude toward doctors
 5. Attitude toward nurses
 6. Attitude toward midwives
 7. Attitude toward alternative medicine
 8. Nurturing personality trait
 9. Dominant personality trait
10. Controlling personality trait

Table 3.7.

Table X

Means With Confidence Intervals and Standard Deviations of Nurses' and Midwives' Work Attitudes and Personality Traits

Variable	Nurses[a]			Midwives[b]		
	M	95% CI	*SD*	*M*	95% CI	*SD*
Satisfaction with work	10.3	[9.9, 10.7]	2.4	12.5	[11.8, 13.2]	3.4
Commitment to work	24.5	[23.7, 25.3]	4.5	25.2	[24.5, 25.9]	3.1
Organizational citizenship behaviors	8.9	[8.7, 9.1]	1.2	8.7	[8.3, 9.1]	2.0
Attitude toward doctors	10.5	[10.0, 11.0]	2.8	8.1	[7.8, 8.4]	1.2
Attitude toward nurses	13.2	[12.9, 13.5]	1.9	13.5	[13.1, 13.8]	1.6
Attitude toward midwives	9.5	[9.3, 9.8]	1.4	14.9	[14.6, 15.2]	1.6
Attitude toward alternative medicine	9.3	[8.8, 9.8]	2.7	11.5	[11.2, 11.8]	1.2
Nurturing personality trait	24.6	[23.9, 25.3]	4.0	23.2	[22.4, 24.0]	3.5
Dominant personality trait	20.1	[19.5, 20.7]	3.5	23.1	[22.5, 23.7]	2.8
Controlling personality trait	18.3	[17.8, 18.8]	2.8	21.8	[21.0, 22.6]	3.4

Note. CI = confidence interval.
[a]*n* = 124. [b]*n* = 79.

▇ Table 3.8.

Table X

Confidence intervals are presented in a different location here than in Table 3.7.

Means With Confidence Intervals (CIs) and Standard Deviations of Nurses' and Midwives' Work Attitudes and Personality Traits

	Nurses[a]		Midwives[b]	
Variable	*M (SD)*	95% CI	*M (SD)*	95% CI
Satisfaction with work	10.3 (2.4)	[9.9, 10.7]	12.5 (3.4)	[11.8, 13.2]
Commitment to work	24.5 (4.5)	[23.7, 25.3]	25.2 (3.1)	[24.5, 25.9]
Organizational citizenship behaviors	8.9 (1.2)	[8.7, 9.1]	8.7 (2.0)	[8.3, 9.1]
Attitude toward doctors	10.5 (2.8)	[10.0, 11.0]	8.1 (1.2)	[7.8, 8.4]
Attitude toward nurses	13.2 (1.9)	[12.9, 13.5]	13.5 (1.6)	[13.1, 13.8]
Attitude toward midwives	9.5 (1.4)	[9.3, 9.8]	14.9 (1.6)	[14.6, 15.2]
Attitude toward alternative medicine	9.3 (2.7)	[8.8, 9.8]	11.5 (1.2)	[11.2, 11.8]
Nurturing personality trait	24.6 (4.0)	[23.9, 25.3]	23.2 (3.5)	[22.4, 24.0]
Dominant personality trait	20.1 (3.5)	[19.5, 20.7]	23.1 (2.8)	[22.5, 23.7]
Controlling personality trait	18.3 (2.8)	[17.8, 18.8]	21.8 (3.4)	[21.0, 22.6]

[a]*n* = 124. [b]*n* = 79.

Standard deviations are in parentheses so they may be easily identified in the table. If means and standard deviations were in separate columns, parentheses would not be used. See Table 3.9.

Table 3.9.

> The lower limits and the upper limits of confidence intervals are frequently labeled in a table.

Table X

Means With Confidence Intervals and Standard Deviations of Nurses' and Midwives' Work Attitudes and Personality Traits

Variable	Nurses[a]				Midwives[b]			
			95% CI				95% CI	
	M	*SD*	*LL*	*UL*	*M*	*SD*	*LL*	*UL*
Satisfaction with work	10.3	2.4	9.9	10.7	12.5	3.4	11.8	13.2
Commitment to work	24.5	4.5	23.7	25.3	25.2	3.1	24.5	25.9
Organizational citizenship behaviors	8.9	1.2	8.7	9.1	8.7	2.0	8.3	9.1
Attitude toward doctors	10.5	2.8	10.0	11.0	8.1	1.2	7.8	8.4
Attitude toward nurses	13.2	1.9	12.9	13.5	13.5	1.6	13.1	13.8
Attitude toward midwives	9.5	1.4	9.3	9.8	14.9	1.6	14.6	15.2
Attitude toward alternative medicine	9.3	2.7	8.8	9.8	11.5	1.2	11.2	11.8
Nurturing personality trait	24.6	4.0	23.9	25.3	23.2	3.5	22.4	24.0
Dominant personality trait	20.1	3.5	19.5	20.7	23.1	2.8	22.5	23.7
Controlling personality trait	18.3	2.8	17.8	18.8	21.8	3.4	21.0	22.6

Note. CI = confidence interval; *LL* = lower limit; *UL* = upper limit.
[a]*n* = 124. [b]*n* = 79.

> Means and standard deviations may be placed in separate columns.

4

Chi-Square

What Is It?

There are several different *chi-square tests,* statistical procedures in which the results are evaluated in terms of the chi-square distribution. In this chapter we focus on the chi-square test of independence, in which one examines whether there is an association between two categorical variables.

What Tables Are Used?

The tables that are commonly used to present the results of chi-square analyses are frequency tables with a column for chi-square values to indicate whether certain frequencies are significantly different from each other as a function of another variable (Tables 4.1 and 4.2).

"Play It Safe" Table

Table 4.1 is the "safe" choice because it follows the recommendation to report exact probabilities in the sixth edition of the *Publication Manual of the American Psychological Association.*

Example 4.1

In this study, the researchers were investigating certain health problems in infancy. They were interested in whether there is any sex difference in the rate of occurrence of these health problems. Their data came from the health records of the 1st year of life for 266 infants who received regular medical care and immunizations. The independent variable is the type of health problem: eye infection, ear infection, strep throat infection, upper respiratory virus, pneumonia, and bronchitis. The dependent variable is the frequency of occurrence of these types of illness.

Variables for Example 4.1

Independent Variables

1. Type of illness (eye infection, ear infection, strep throat infection, upper respiratory virus, pneumonia, bronchitis)
2. Sex (boys vs. girls)

Dependent Variable

1. Frequency of occurrence of illness (occurrence vs. nonoccurrence)

Table 4.1.

This is the "Play It Safe" table.

Note that according to APA Style language guidelines, *male* and *female* should be used only as adjectives, not nouns, when they refer to human beings.

Table X

Prevalence of Six Illnesses in Male Infants (n = 123) and Female Infants (n = 143) in the 1st Year of Life

Illness	Male infants		Female infants		$\chi^2(1)$	p
	n	%	n	%		
Eye infection	46	37	27	19	11.39	.001
Ear infection	73	59	53	37	13.17	<.001
Strep throat infection	26	21	31	22	0.01	.915
Upper respiratory virus	96	78	68	48	26.01	<.001
Pneumonia	13	11	23	16	1.72	.186
Bronchitis	14	11	6	4	4.91	.028

■ **Table 4.2.**

> This table presents the same information as Table 4.1 but uses a different method for presenting significance levels.

Table X

Prevalence of Six Illnesses in Male Infants (n = 123) and Female Infants (n = 143) in the 1st Year of Life

Illness	Male infants		Female infants		
	n	%	n	%	$\chi^2(1)$
Eye infection	46	37	27	19	11.39**
Ear infection	73	59	53	37	13.17***
Strep throat infection	26	21	31	22	0.01
Upper respiratory virus	96	78	68	48	26.01***
Pneumonia	13	11	23	16	1.72
Bronchitis	14	11	6	4	4.91*

*$p < .05$. **$p < .01$. ***$p < .001$.

t Test of Means

What Is It?

A *t* test is used to compare mean values on a continuous variable—for instance, to determine whether two means from the same population differ from each other (within subject or paired-samples *t* test) or whether two means from different populations differ from each other (between-subjects or independent-samples *t* test).

What Tables Are Used?

When there is only one *t* test to report, no table is required—the results are presented in the text rather than in a table. If results of several *t* tests are to be presented, then a table could be created. Tables 5.1 and 5.2 illustrate the results of multiple separate *t* tests.

"Play It Safe" Table

There is no comprehensive table for a single *t* test. Usually the means, standard deviations (or standard errors), *t* values, degrees of freedom, significance levels, and effect sizes are reported within the text. However, a table can be presented for the results of several *t* tests. In this case, the most comprehensive table would include all of these statistics. Table 5.1 would be used if the degrees of freedom for each *t* test were the same, and Table 5.2 would be used if the degrees of freedom for each *t* test differed. Results for within-subject *t* tests and between-subjects *t* tests are similarly presented.

Example 5.1

A researcher wished to determine the effectiveness of a new mnemonic task on four different types of memory: memory for lists of faces, memory for lists of words, memory for lists of nonsense words, and memory for lists of two-digit numbers. The researcher conducted four experiments, one for each type of memory. Each study consisted of a control group and an experimental group, with 10 participants in each group. The experimental group received training on the mnemonic task for 2 hr, whereas the control group received no memory training.

There is one independent variable, the mnemonic task. There are four dependent variables: memory for lists of faces, memory for lists of words, memory for lists of nonsense words, and memory for lists of two-digit numbers.

Variables for Example 5.1

Independent Variable

1. Mnemonic task (program provided or no program provided)

Dependent Variables

1. Memory for lists of faces
2. Memory for lists of words
3. Memory for lists of nonsense words
4. Memory for lists of two-digit numbers

■ **Table 5.1.**

This is the "Play It Safe" table for presenting the results of more than one *t* test.

Table X

Group Differences for Memory Tasks Between Groups That Did or Did Not Learn a New Mnemonic Task

Memory measure	No mnemonic task M	SD	Mnemonic task M	SD	t(18)	p	Cohen's d
Faces	8.90	2.02	12.20	2.66	−3.12	.006	1.40
Words	6.70	2.21	10.00	2.98	−2.81	.012	1.26
Nonsense words	3.80	2.39	4.70	2.31	−0.86	.404	0.38
Two-digit numbers	8.10	2.23	8.20	2.30	−0.10	.923	0.04

Example 5.2

This example study was the same in all respects as that described in Example 5.1 except that in Example 5.2 the number of participants assigned to each condition varied. As such, the degrees of freedom for the *t* tests differ.

Table 5.2.

This is the "Play It Safe" table for presenting the results of more than one *t* test when the number of participants is not the same for each *t* test.

Table X

Memory Differences Between Individuals Who Did Learn a New Mnemonic Task and Those Who Did Not Learn a New Mnemonic Task

Memory measure	No mnemonic task		Mnemonic task		df	t	p	Cohen's d
	M	*SD*	*M*	*SD*				
Faces	9.27	2.28	11.87	2.90	17	−2.19	.043	1.00
Words	7.58	2.99	9.38	2.72	18	−1.36	.192	0.63
Nonsense words	5.54	4.68	4.78	2.44	20	0.45	.661	0.20
Two-digit numbers	8.22	2.33	8.09	2.21	18	0.13	.899	0.06

6

Post Hoc and A Priori
Tests of Means

What Is It?

Post hoc and a priori analyses are used to compare specific group means in studies where the independent variables have more than two levels. Post hoc comparison techniques (e.g., Tukey's honestly significant difference and Scheffé) are often used after an overall significant test has been obtained for testing hypotheses or for exploring unhypothesized relations. A priori (or planned) comparisons are planned before data are analyzed; theory or prior empirical evidence is often used as a guide in determining which comparisons are to be made.

What Tables Are Used?

Often, the results of post hoc or a priori analyses are not presented in a table. If several of these analyses are conducted, however, then a table can be a useful way of summarizing the results. Generally, significant post hoc or a priori analyses are indicated in a table of means and standard deviations (Table 6.1 or 6.2).

"Play It Safe" Table

The "safe" choice for a post hoc or a priori table is Table 6.1. Although Tables 6.1 and 6.2 are equally comprehensive, Table 6.1 is the more conventional format for presenting post hoc or a priori analysis results.

Example 6.1

In this study, the researchers were interested in the creativity of severely depressed versus mildly depressed versus individuals who are not depressed. A group of 60 severely depressed individuals was recruited from a psychiatric hospital. A group of 60 mildly depressed individuals was also recruited through the hospital. A third group of 60 individuals who had never been depressed was recruited using newspaper advertisements. All three groups of participants were assessed for creativity using four different measures: the Franklin Creativity Test, a rating of creative accomplishments, a creative writing exercise, and a peer rating of creativity. The researchers were interested in whether the three groups of participants differed significantly on any of these measures. In this study, the independent variable is participant group (severely depressed vs. mildly depressed vs. not depressed). The dependent variables are the four measures of creativity.

Variables for Example 6.1

Independent Variable

1. Participant group (severely depressed vs. mildly depressed vs. not depressed)

Dependent Variables

1. Franklin Creativity Test
2. Creative accomplishments
3. Creative writing exercise
4. Peer rating

▨ Table 6.1.

This is the "Play It Safe" table for a post hoc or a priori analysis.

The sample size for each participant group could be presented under each group name, particularly if group sizes vary.

Table X

Mean Scores on Four Measures of Creativity as a Function of Participant Group

| | Participant group | | | | | |
| | Severely depressed | | Mildly depressed | | Not depressed | |
Creativity measure	*M*	*SD*	*M*	*SD*	*M*	*SD*
Franklin Creativity Test	12.8$_a$	6.7	10.2$_b$	5.8	6.3$_{a,b}$	4.4
Accomplishments	21.2$_a$	11.2	23.1$_b$	12.2	13.1$_{a,b}$	7.7
Writing exercise	8.7$_a$	3.3	6.7$_a$	2.9	5.5$_a$	3.3
Peer rating	9.1	3.4	9.0	4.1	8.0	3.1

Note. Means in a row sharing subscripts are significantly different from each other. For all measures, higher means indicate higher creativity scores.

When using subscripts to denote significant contrasts, it is important to include in the table note that means in a row with the same subscript (e.g., a) have been found to be significantly different.

Different sets of subscripts can be used for each row in the table (i.e., a,b for row 1; c,d for row 2, etc).

■ **Table 6.2.**

Table X

Creativity in Individuals With Severe Depression, With Mild Depression, and With No Depression

Creativity measure	Severely depressed (1)		Mildly depressed (2)		Not depressed (3)		Post hoc
	M	SD	M	SD	M	SD	
Franklin Creativity Test	12.8	6.7	10.2	5.8	6.3	4.4	3 < 1, 2
Accomplishments	21.2	11.2	23.1	12.2	13.1	7.7	3 < 1, 2
Writing exercise	8.7	3.3	6.7	2.9	5.5	3.3	3 < 2 < 1
Peer rating	9.1	3.4	9.0	4.1	8.0	3.1	3 = 2 = 1

Note. The numbers in parentheses in column heads refer to the numbers used for illustrating significant differences in the "Post hoc" column.

7

Correlation

What Is It?

A *correlation* is a measure of the direction and magnitude of the linear relation between two variables. The analysis produces a correlation coefficient, which can range from −1 to 1. The closer the correlation coefficient is to positive or negative 1, the stronger the relationship between the two variables in the analysis.

What Tables Are Used?

When reporting the results of correlational analyses, a table usually is necessary only if there are more than two variables. In that case, the relevant tables are a table of means and standard deviations (or standard errors; Table 7.1) and a table of intercorrelations among all variables included in the analysis (Table 7.2, 7.3, 7.4, or 7.5) or a table of correlations between two sets of variables (Table 7.7). The table of means and standard deviations and the table of intercorrelations can be combined into one table (Table 7.6 or 7.7). The table format used depends on the nature of the study.

"Play It Safe" Table

All of the tables in this chapter could be considered "safe" choices because they all present the same amount of information about the correlations obtained from the data. Thus, no one table is any more or less comprehensive than any other.

Example 7.1

In the study presented here, the researchers constructed a new scale to measure need for achievement. They called their new scale the Dimensions of Achievement Scale (DAS). To examine how the DAS relates to existing measures of need for achievement, the researchers wanted to correlate scores on the DAS with scores on other measures. They had 100 participants complete each of the six measures. Their variables are the different measures of need for achievement: the DAS, the Brunswick Achievement Measure, the Need for Achievement Inventory, the Achievement Perception Test, a peer rating of need for achievement, and a self-rating of need for achievement.

Variables for Example 7.1

1. Dimensions of Achievement Scale
2. Brunswick Achievement Measure
3. Need for Achievement Inventory
4. Achievement Perception Test
5. Peer rating of need for achievement
6. Self-rating of need for achievement

See Chapter 3 for other examples of format for tables of means and standard deviations.

Table 7.1.

Table X

Means and Standard Deviations for Six Measures of Need for Achievement

Measure	M	SD
Dimensions of Achievement Scale	43.21	14.34
Brunswick Achievement Measure	22.22	8.75
Need for Achievement Inventory	12.15	3.47
Achievement Perception Test	14.09	5.37
Peer rating of need for achievement	12.30	5.57
Self-rating of need for achievement	11.91	4.91

Table 7.2.

This is the most common format for a table of intercorrelations.

Table X

Intercorrelations for Dimensions of Achievement Scale and Five Other Need-for-Achievement Measures

Measure	1	2	3	4	5	6
1. Dimensions of Achievement Scale	—					
2. Brunswick Achievement Measure	.76	—				
3. Need for Achievement Inventory	.70	.88	—			
4. Achievement Perception Test	.56	.65	.61	—		
5. Peer rating of need for achievement	.45	.55	.52	.67	—	
6. Self-rating of need for achievement	.53	.56	.43	.37	.87	—

Note. All coefficients are significant at $p < .01$.

Table 7.3.

In this table, the data for two groups of participants are presented—one above the diagonal and one below the diagonal.

Table X

Intercorrelations for Scores on Six Measures of Need for Achievement as a Function of Gender

Measure	1	2	3	4	5	6
1. Dimensions of Achievement Scale	—	.86	.76	.60	.43	.63
2. Brunswick Achievement Measure	.66	—	.80	.70	.55	.50
3. Need for Achievement Inventory	.64	.96	—	.62	.52	.40
4. Achievement Perception Test	.52	.60	.60	—	.77	.37
5. Peer rating of need for achievement	.47	.55	.52	.57	—	.90
6. Self-rating of need for achievement	.43	.61	.45	.37	.85	—

Note. Intercorrelations for male participants ($n = 50$) are presented above the diagonal, and intercorrelations for female participants ($n = 50$) are presented below the diagonal. All coefficients are significant at $p < .01$.

■ **Table 7.4.**

Where relevant, researchers sometimes present coefficient alphas for the different measures in a correlation table, as illustrated in this table.

Table X

Intercorrelations and Coefficient Alphas for Scores on Six Measures of Need for Achievement

Measure	1	2	3	4	5	6
1. Dimensions of Achievement Scale	(.91)					
2. Brunswick Achievement Measure	.76	(.89)				
3. Need for Achievement Inventory	.70	.88	(.92)			
4. Achievement Perception Test	.56	.65	.61	(.84)		
5. Peer rating of need for achievement	.45	.55	.52	.67	(.78)	
6. Self-rating of need for achievement	.53	.56	.43	.37	.87	(.80)

Note. Coefficient alphas are presented in parentheses along the diagonal. All coefficients are significant at $p < .01$.

■ **Table 7.5.**

Researchers sometimes include 95% confidence intervals for correlation coefficients, as in this example.

Table X

Intercorrelations Among Six Measures of Need for Achievement

Measure	1	2	3	4	5	6
1. DAS	—					
2. BAM	.76 [.55, .97]	—				
3. NAchI	.70 [.48, .92]	.88 [.72, .99]	—			
4. APT	.56 [.33, .89]	.65 [.40, .90]	.61 [.36, .86]	—		
5. Peer rating	.45 [.19, .71]	.55 [.28, .82]	.52 [.25, .79]	.67 [.43, .91]	—	
6. Self-rating	.53 [.27, .79]	.56 [.30, .82]	.43 [.13, .73]	.37 [.06, .68]	.87 [.72, .99]	—

Note. Numbers in brackets are 95% confidence intervals of the correlation coefficients. All coefficients are significant at $p < .01$. DAS = Dimensions of Achievement Scale; BAM = Brunswick Achievement Measure; NAchI = Need for Achievement Inventory; APT = Achievement Perception Test; Peer rating = peer rating of need for achievement; Self-rating = self-rating of need for achievement.

p values fall at the end of a note only if they are represented by asterisks.

Table 7.6.

Instead of using a separate table of means and standard deviations, these values can be included in a correlation table.

Table X

Means, Standard Deviations, and Intercorrelations for Scores on Six Measures of Need for Achievement

Measure	M	SD	1	2	3	4	5	6
1. DAS	43.21	14.34	—					
2. BAM	22.22	8.75	.76	—				
3. NAchI	12.15	3.47	.70	.88	—			
4. APT	14.09	5.37	.56	.65	.61	—		
5. Peer rating	12.30	5.57	.45	.55	.52	.67	—	
6. Self-rating	11.91	4.91	.53	.56	.43	.37	.87	—

Note. All coefficients are significant at $p < .01$. DAS = Dimensions of Achievement Scale; BAM = Brunswick Achievement Measure; NAchI = Need for Achievement Inventory; APT = Achievement Perception Test; Peer rating = peer rating of need for achievement; Self-rating = self-rating of need for achievement.

Example 7.2

There are six subscales of the DAS. Each subscale measures a different domain in which need for achievement might be manifested: School, Family Relationships, Friendships, Work, Athletics, and Hobbies. The researchers wanted to determine how each of the DAS subscales correlated with the other measures of achievement included in Example 7.1 (see Table 7.7).

Table 7.7.

Table X

Means, Standard Deviations, and Correlations of Dimensions of Achievement Scale Subscales With Measures of Need for Achievement

			Measure				
DAS subscale	*M*	*SD*	BAM	NAchl	APT	Peer	Self
School	22.3	7.7	.76**	.89**	.45*	.34*	.55*
Family Relationships	20.9	6.7	.86**	.79**	.73**	.44*	−.20
Friendships	16.8	6.1	.77**	.82**	.23	.39*	.45*
Work	24.6	6.9	.80**	.79**	.45*	.41*	.63*
Athletics	22.5	9.8	.66*	.77**	−.19*	.75**	−.21*
Hobbies	14.3	9.1	.79**	.88**	.56*	.21	.22

Note. DAS = Dimensions of Achievement Scale; BAM = Brunswick Achievement Measure; NAchl = Need for Achievement Inventory; APT = Achievement Perception Test; Peer = peer rating of need for achievement; Self = self-rating of need for achievement.
*$p < .05$. **$p < .01$.

Means and standard deviations could be presented in a separate table. See Chapter 3 for other examples of format for tables of means and standard deviations.

8

Canonical Correlation

What Is It?

The *canonical correlation* is used to measure the linear relationships between two sets of variables. For each set of variables, the analysis computes a new variable (a *canonical variate* or *root*) made from linear combinations of the original variables, such that the correlation of the original and new variables is maximized.

What Tables Are Used?

The results of canonical correlations usually are reported in one table that includes the correlations and standardized canonical coefficients between the sets of variables and their canonical variates (see Table 8.1 or 8.2).

"Play It Safe" Table

The "Play It Safe" choice for the format of the canonical correlation table is Table 8.1 because it is most comprehensive.

Example 8.1

The sample data used here are from a study in which the researchers were interested in the relationship between two sets of variables. One set of variables measured job satisfaction, and the other set measured participants' personal characteristics. The job satisfaction variables included an overall satisfaction rating, satisfaction with working conditions, satisfaction with amount of work, and satisfaction with promotion prospects. The personal characteristics variables included education, health, income, and age. The researchers wanted to determine how these two sets of variables were related for a sample of employees at a major multinational corporation.

The researchers should report the significant canonical correlations in the text. For this example, there are two pairs of canonical variates that account for significant relationships between the two sets of variables. These are the canonical variates that should be presented in the table.

Variables for Example 8.1

1. Job satisfaction (overall satisfaction rating, satisfaction with working conditions, satisfaction with amount of work, satisfaction with promotion prospects)
2. Personal characteristics (education, health, income, age)

 **Table 8.1.**

This is the "Play It Safe" table for canonical correlation results.

Table X

Correlations and Standardized Canonical Coefficients Between Job Satisfaction and Personal Characteristics Variables and Their Canonical Variates

	First variate		Second variate	
Variable	Correlation	Canonical coefficient	Correlation	Canonical coefficient
Job satisfaction				
Overall satisfaction rating	.72	.66	.25	.59
Working conditions	−.56	−.34	−.31	−.44
Amount of work	.78	.44	.12	.23
Promotion prospects	.19	.12	.45	.34
Personal characteristics				
Education	.87	.76	.45	.23
Health	.67	.52	.21	.11
Income	.92	.69	.59	.54
Age	−.65	−.55	−.33	−.08

Table 8.2.

Table X

Canonical Analysis of Job Satisfaction and Personal Characteristics Variables

Variable	Standardized canonical coefficient	
	Root 1	Root 2
Job satisfaction		
Overall satisfaction rating	.66	.59
Working conditions	−.34	−.44
Amount of work	.44	.23
Promotion prospects	.12	.34
Personal characteristics		
Education	.76	.23
Health	.52	.11
Income	.69	.54
Age	−.55	−.08

9

Analysis of Variance

One-Way Analysis of Variance: What Is It?

The one-way analysis of variance (ANOVA) is used to assess the differences between two or more group means when there is one independent variable and one dependent variable. Here the completely randomized design will be illustrated where each case or participant is represented in only one cell.

What Tables Are Used?

When there is only one analysis to report, no table is required (see section 5.03 of the *Publication Manual of the American Psychological Association*). The means, standard deviations, and ANOVA results are presented in the text rather than in a table. The results of a single ANOVA may be presented in tabular form in theses and reports, but this would have to be verified by the writer. In addition, one-way ANOVA results often are presented in tables for theses and for journals if there are several independent ANOVAs to be reported; however, some disciplines prefer not to present ANOVA results in tables but rather to present relevant means in tables or in figures. The following two examples illustrate what is commonly presented in one-way ANOVA tables. For Example 9.1, the results of a single ANOVA are illustrated in Table 9.1. For Example 9.2, the results of multiple separate ANOVAs are presented (Tables 9.2–9.6).

"Play It Safe" Table

The most comprehensive ANOVA table includes the degrees of freedom, sums of squares, mean squares, F ratios, p values, and effect sizes (e.g., η^2 or ω^2; see Table 9.1 for a single ANOVA and Tables 9.3 and 9.4 for several independent ANOVAs). In addition, means and standard deviations are presented in the text for the results of a single ANOVA or in a table if several independent ANOVAs are presented (see Table 9.2). Note that the "Play It Safe" tables presented here are not what are usually presented in journals. As stated in the previous section "What Tables Are Used?", when preparing a manuscript for a journal, present single ANOVA results in the text (degrees of freedom, F ratio, p value, and effect size) and not in a table; for multiple ANOVA results, present the means and standard deviations either in a table or in a figure, and if presenting the means and standard deviations in a table, include degrees of freedom, F ratios, p values, and effect sizes (see Table 9.5).

Example 9.1

A company wished to see the effects of three types of training programs on employees' job performance 3 months after the programs had been completed. There are three levels of the independent variable (i.e., three different training programs):

1. *Coworker program.* Employee is taught for 3 days by an experienced coworker and is provided with an information manual.
2. *Consultant program.* Employee is taught for 3 days by an external consultant and is provided with an information manual.
3. *Self-program.* Employee is provided with an information manual and learns on his or her own for 3 days.

The dependent variable is job performance. This was measured using a seven-item scale completed by each employee's supervisor. There were 40 employees in each of the three training programs.

To summarize, the study consisted of one independent variable (training program) and one dependent variable (job performance).

> ### Variables for Example 9.1
>
> **Independent Variable**
>
> 1. Training program (coworker, consultant, or self)
>
> **Dependent Variable**
>
> 1. Job performance

The results of an ANOVA conducted on the data from the study described in this example are presented in Table 9.1. The means and standard deviations for the three training programs could be provided in the text; although presented here, a table is not necessary to display these simple results.

If there is more than one one-way ANOVA to be reported, one table can be used to present the descriptive statistics (e.g., means and standard deviations), and a separate table can be used to present the ANOVA results. However, most journals prefer that authors combine the descriptive statistics with the ANOVA results.

■ **Table 9.1.**

This is the "Play It Safe" table for a single ANOVA for theses and reports but is not the convention for published manuscripts.

Table X

One-Way Analysis of Variance Summary Table for the Effects of Training Program on Job Performance

Source	df	SS	MS	F	p	η^2
Between-group	2	53.60	26.80	5.79	.004	.09
Within-group	117	541.20	4.63			
Total	119	594.80				

For journals, the results of a single ANOVA are normally reported in text form only.

Example 9.2

In this example, a company wished to see the effects of three types of training programs on numerous attitudes and behaviors 6 months after the programs had been completed. Again, there are three levels of the one independent variable: coworker program, consultant program, and self-program. There were 100 employees in each of the three training programs (for a total of 300 employees).

There are seven dependent variables (all continuous variables): (a) job performance, (b) organizational commitment, (c) job commitment, (d) job satisfaction, (e) turnover intention, (f) job stress, and (g) role ambiguity. Because there are seven dependent variables, seven separate analyses must be conducted.

Variables for Example 9.2

Independent Variable

1. Training program (coworker, consultant, or self)

Dependent Variables

1. Job performance
2. Organizational commitment
3. Job commitment
4. Job satisfaction
5. Turnover intention
6. Job stress
7. Role ambiguity

For this example, the descriptive statistics could be presented in a table such as Table 9.2. (Note that other examples of format for tables of means and standard deviations can be found in Chapter 3.)

The results of the ANOVAs for the seven separate analyses could be presented in a number of ways. Tables 9.3 and 9.4 present, for all of the analyses, the degrees of freedom, sums of squares, and mean squares for both between-groups and within-groups calculations. Tables 9.5 and 9.6 are examples of how descriptive statistics and ANOVA F ratios can be combined in the same table. Also presented are degrees of freedom, F ratios, p values, and effect sizes (η^2).

Table 9.2.

See Chapter 3 for other examples of format for tables of means and standard deviations.

Table X

Means and Standard Deviations for Three Training Programs and Seven Dependent Variables

	Coworker		Consultant		Self	
Variable	*M*	*SD*	*M*	*SD*	*M*	*SD*
Job performance	12.34	2.89	11.78	3.45	10.90	2.98
Organizational commitment	9.54	1.51	9.67	1.47	9.89	1.32
Job commitment	3.35	0.89	3.41	0.96	3.33	0.82
Job satisfaction	5.67	1.01	4.79	0.99	3.45	1.10
Turnover intention	1.44	0.56	1.89	0.67	2.02	0.59
Job stress	15.87	3.56	15.32	3.24	17.04	3.18
Role ambiguity	4.45	1.32	4.39	4.04	1.25	1.35

■■ **Table 9.3.**

This table, along with a table of means and standard deviations, is a "Play It Safe" table for several one-way ANOVAs for theses and reports but is not the convention for published manuscripts (see Table 9.5).

Because the degrees of freedom do not vary in this table, they are presented in parentheses following the *F* heading. They may also be presented as a lettered footnote to the table (see Table 9.12). When the degrees of freedom are not the same for each analysis, they can be presented in a separate *df* column (see Table 9.1).

Table X

One-Way Analysis of Variance for the Effects of Coworker, Consultant, and Self-Training Programs on Seven Dependent Variables

Variable and source	SS	MS	F(2, 297)	p	η^2
Job performance					
Between	105.39	52.69	5.43	.005	.04
Within	2,884.36	9.71			
Organizational commitment					
Between	6.26	3.13	1.51	.221	.01
Within	612.16	2.06			
Job commitment					
Between	0.35	0.17	0.22	.804	.00
Within	236.22	0.80			
Job satisfaction					
Between	249.95	124.97	116.79	<.001	.44
Within	317.81	1.07			
Turnover intention					
Between	18.53	9.26	25.02	<.001	.14
Within	109.95	0.37			
Job stress					
Between	154.33	77.16	6.96	.001	.04
Within	3,295.07	11.10			
Role ambiguity					
Between	670.11	335.05	50.54	<.001	.25
Within	1,968.76	6.63			

■ **Table 9.4.**

This table, along with a table of means and standard deviations, is a "Play It Safe" table for several one-way ANOVAs for theses and reports but is not the convention for published manuscripts (see Table 9.5).

The Effects of Training Programs on Job Performance, Organizational Commitment, Job Commitment, Job Satisfaction, Turnover Intention, Job Stress, and Role Ambiguity

Variable and source	SS	MS	$F(2, 297)$	p	η^2
Job performance					
Between	105.39	52.69	5.43	.005	.04
Error	2,884.36	9.71			
Organizational commitment					
Between	6.26	3.13	1.51	.221	.01
Error	612.16	2.06			
Job commitment					
Between	0.35	0.17	0.22	.804	.00
Error	236.22	0.80			
Job satisfaction					
Between	249.95	124.97	116.79	<.001	.44
Error	317.81	1.07			
Turnover intention					
Between	18.53	9.26	25.02	<.001	.14
Error	109.95	0.37			
Job stress					
Between	154.33	77.16	6.96	.001	.04
Error	3,295.07	11.10			
Role ambiguity					
Between	670.11	335.05	50.54	<.001	.25
Error	1,968.76	6.63			

Table 9.5.

This style of table is what is more commonly found in journals.

Table X

Means, Standard Deviations, and One-Way Analyses of Variance for the Effects of Coworker, Consultant, and Self-Training Programs on Seven Dependent Variables

Variable	Coworker		Consultant		Self		$F(2, 297)$	p	η^2
	M	SD	M	SD	M	SD			
Job performance	12.34	2.89	11.78	3.45	10.90	2.98	5.43	.005	.04
Organizational commitment	9.54	1.51	9.67	1.47	9.89	1.32	1.51	.221	.01
Job commitment	3.35	0.89	3.41	0.96	3.33	0.82	0.22	.804	.00
Job satisfaction	5.67	1.01	4.79	0.99	3.45	1.10	116.79	<.001	.44
Turnover intention	1.44	0.56	1.89	0.67	2.02	0.59	25.02	<.001	.14
Job stress	15.87	3.56	15.32	3.24	17.04	3.18	6.96	.001	.04
Role ambiguity	4.45	1.32	4.39	4.04	1.25	1.35	50.54	<.001	.25

Because the degrees of freedom are the same here, they are presented in parentheses following the *F* heading. If column spacing is tight, the degrees of freedom may be presented as a lettered footnote (see Table 9.12).

Table 9.6.

Table X

Means, Standard Deviations, and One-Way Analyses of Variance for the Effects of Coworker, Consultant, and Self-Training Programs on Seven Dependent Variables

	Coworker		Consultant		Self			
Variable	*M*	*SD*	*M*	*SD*	*M*	*SD*	$F(2, 297)$	η^2
Job performance	12.34	2.89	11.78	3.45	10.90	2.98	5.43*	.04
Organizational commitment	9.54	1.51	9.67	1.47	9.89	1.32	1.51	.01
Job commitment	3.35	0.89	3.41	0.96	3.33	0.82	0.22	.00
Job satisfaction	5.67	1.01	4.79	0.99	3.45	1.10	116.79**	.44
Turnover intention	1.44	0.56	1.89	0.67	2.02	0.59	25.02**	.14
Job stress	15.87	3.56	15.32	3.24	17.04	3.18	6.96**	.04
Role ambiguity	4.45	1.32	4.39	4.04	1.25	1.35	50.54**	.25

*$p < .01$. **$p < .001$.

In this sample table, asterisk notes are used to express significance levels. Note, however, that APA Style recommends reporting exact significance levels.

Factorial Analysis of Variance: What Is It?

The *factorial* ANOVA (or *multifactor* or *multiway* ANOVA) is similar to the one-way ANOVA except there are two or more independent variables. The effects of the independent variables on a single dependent variable are examined.

What Tables Are Used?

Two types of tables usually are used: (a) a table that identifies the means and standard deviations for each cell of the design (see Tables 9.7 and 9.15) and (b) an ANOVA summary table (see Tables 9.8, 9.9, 9.11, 9.12, 9.13, 9.14, and 9.16). In journals, descriptive statistics and ANOVA results are often presented together (see Table 9.10) or the ANOVA results are presented in the text only (marginal means rather than cell means are often presented when there are no significant interaction effects).

What information the writer presents in the ANOVA table (e.g., degrees of freedom, sums of squares, mean squares, *F* ratios, significance levels, or effect sizes) depends on what guidelines are being followed (e.g., report, thesis, journal), space availability, how comprehensive he or she wishes to be, and whether the results of more than one ANOVA are to be presented in the same table. Examples 9.3 and 9.5 provide sample tables for the results of a single factorial ANOVA (Tables 9.7, 9.8, 9.9, and Tables 9.15

and 9.16, respectively). Example 9.4 provides sample tables for the presentation of more than one factorial ANOVA in a single table (Tables 9.10, 9.11, 9.12, 9.13, and 9.14). Note that researchers often will present significant interactions in a figure (this is not illustrated in this book but can be found in *Displaying Your Findings: A Practical Guide for Creating Figures, Posters, and Presentations, Sixth Edition*).[1]

"Play It Safe" Table

The most comprehensive table includes the degrees of freedom, sums of squares, mean squares, F ratios, significance levels for the main effects and interaction effects, and effect sizes. Information regarding the within-groups, between-groups, and total sources of variation is included as well (see Table 9.8). If the goal is to be comprehensive in presenting the results, then a table of means and standard deviations also should be included (see Table 9.7). These two types of tables (table of means, ANOVA summary table) are typically found in theses and reports. However, many published articles will present only a table of means and will present the F ratios, degrees of freedom, p values, and effect sizes in the text.

Example 9.3

A school wanted to see the effects of different classroom and laboratory instructional media on students' grades. A total of 120 students attended a 3-hr psychology class once a week for 3 months. There are three levels of this independent variable (i.e., three different classroom conditions):

1. *Lecture only.* An instructor lectured on material obtained from a textbook, but students were not provided with the textbook.
2. *Textbook only.* Students were provided with only a textbook that they could read during the classroom period.
3. *Multimedia computer-assisted instruction only.* Students learned about the textbook material through a multimedia computer-assisted instructional program during the classroom period; no textbook was provided.

In addition, students attended one of two 1-hr laboratory instruction sessions that took place once a week. The content of the laboratory sessions focused on material presented in the previous class. There are two levels of this independent variable (i.e., two different laboratory conditions):

1. *Group discussion only.* Students were involved in group discussion sessions in which the instructor acted as a facilitator.
2. *Problem solving only.* Students were involved in problem-solving sessions.

Therefore, there are two independent variables (three levels of classroom instruction and two levels of laboratory instruction) and one dependent variable (overall course grade, measured here as a continuous variable).

[1]Nicol, A. A. M., & Pexman, P. M. (2010). *Displaying your findings: A practical guide for creating figures, posters, and presentations* (6th ed.). Washington, DC: American Psychological Association.

Variables for Example 9.3

Independent Variables

1. Classroom instruction (lecture, textbook, or multimedia computer-assisted instruction)
2. Laboratory instruction (group discussion or problem solving)

Dependent Variable

1. Overall course grade

Table 9.7.

Descriptive statistics such as means and standard deviations often are presented first.

See Chapter 3 for other examples of format for tables of means and standard deviations.

Table X

Means and Standard Deviations for Class Conditions as a Function of Laboratory Instruction Condition

| | Laboratory instruction | | | |
| | Group discussion | | Problem solving | |
Class	*M*	*SD*	*M*	*SD*
Lecture	72.63	2.01	71.18	2.11
Textbook	70.50	1.99	72.19	2.86
Multimedia	79.16	6.29	78.28	4.76

Means are sometimes displayed in figures rather than in a table. This will depend on the discipline or the subject matter.

When ANOVA results are presented, the degrees of freedom, sums of squares, mean squares, F ratios, significance levels, and effect sizes are often presented for the different effects (i.e., main effects and interaction effects). The degrees of freedom, sum of squares, and mean squares are presented for the within-groups source of variance. The degrees of freedom and sums of squares are presented for the between-groups and total sources of variation. Table 9.8 is a sample based on the study described in Example 9.3. To simplify Table 9.8, one could omit information regarding the total sources of variation, as in Table 9.9. ANOVA results for journal articles are frequently presented just in the text (F ratios, degrees of freedom, p values, and effect sizes).

▦ Table 9.8.

This table, along with a table of means and standard deviations, is the "Play It Safe" table for a factorial ANOVA with two independent variables. For journals, the F statistic along with the degrees of freedom, p value, and effect size is frequently presented in the text and not in a table.

Table X

Summary Table for Two-Way Analysis of Variance of the Effects of Class and Laboratory Conditions on Overall Course Grade

Source	df	SS	MS	F	p	η²
Class	2	1,350.11	675.06	48.85	<.001	.462
Laboratory	1	1.36	1.36	0.10	.754	.001
Class × Laboratory	2	56.01	28.00	2.02	.137	.034
Within cells	114	1,575.27	13.82			
Total	120	863.09				

▦ Table 9.9.

Table X

Two-Way Analysis of Variance for Class and Laboratory Conditions on Overall Course Grade

Source	SS	MS	F	p	η²
Class	1,350.11	675.06[a]	48.85	<.001	.462
Laboratory	1.36	1.36[b]	0.10	.754	.001
Class × Laboratory	56.01	28.00[a]	2.02	.137	.034
Within cells	1,575.27	13.82			

[a]$df = 2, 114.$ [b]$df = 1, 114.$

The total source of variation has not been included in this table. The degrees of freedom have been included in a note to save space.

When several factorial ANOVAs have been conducted, the researcher can make a separate ANOVA summary table for each ANOVA, such as Tables 9.8 and 9.9 (this may be recommended for reports and theses). Otherwise, including all of the analyses in a summary table such as Tables 9.10 through 9.14, which do not contain as much information, is an alternative.

Example 9.4

As in Example 9.3, there are two independent variables: three levels of classroom instruction and two levels of laboratory instruction. Additional dependent variables have been included: overall course grade, recall and recognition score, problem-solving score, and course satisfaction ratings. Separate analyses have been conducted for each.

The researcher may choose not to include the error term (i.e., within groups) in the table. In some instances, degrees of freedom may not be included within a table. For example, including the degrees of freedom may occupy too much space in the table, or the degrees of freedom may be the same for each set of independent ANOVA analyses and therefore are redundant if repeated within a table column. In such cases, the degrees of freedom may be identified in a lettered table note (see Table 9.12 for an example).

Variables for Example 9.4

Independent Variables

1. Classroom instruction (lecture, textbook, or multimedia computer-assisted instruction)
2. Laboratory instruction (group discussion or problem solving)

Dependent Variables

1. Overall course grade
2. Recall and recognition score
3. Problem-solving score
4. Course satisfaction ratings

■ **Table 9.10.**

Table X

Descriptive statistics and ANOVA results are presented together in this table. This saves space, but some journals may prefer that more information be included (e.g., degrees of freedom, specific *p* values, and effect sizes).

Means, Standard Deviations, and Analysis of Variance (ANOVA) Results for Overall Course Grade, Recall and Recognition, Problem Solving, and Course Satisfaction as a Function of Class and Laboratory Conditions

Class	Group discussion M	Group discussion SD	Problem solving M	Problem solving SD	ANOVA F Class (C)	ANOVA F Lab (L)	ANOVA F C × L
Overall course grade					48.85*	0.10	2.02
Lecture	72.63	2.01	71.18	2.11			
Textbook	70.50	1.99	72.19	2.86			
Multimedia	79.16	6.29	78.28	4.76			
Recall and recognition score					15.09*	0.18	0.13
Lecture	24.65	1.99	24.85	1.93			
Textbook	22.84	1.13	22.76	1.09			
Multimedia	23.98	1.63	24.24	1.68			
Problem-solving score					13.55*	237.17*	2.45
Lecture	13.54	1.17	16.80	1.60			
Textbook	14.32	1.26	18.82	1.44			
Multimedia	13.53	1.30	17.02	1.17			
Course satisfaction ratings					276.87*	78.23*	784.03*
Lecture	66.15	1.56	45.4	2.25			
Textbook	57.85	2.42	49.20	1.99			
Multimedia	55.57	2.06	74.00	3.04			

*$p < .001$.

Depending on the discipline or the subject matter, means are sometimes presented in figures rather than in a table.

Table 9.11.

Only ANOVA results are presented in this table. Degrees of freedom, *F* values, *p* values, and effect sizes are typically presented in the text of a journal article.

Table X

Two-Way Analyses of Variance for Overall Grade, Recall and Recognition and Problem-Solving Scores, and Course Satisfaction Ratings as a Function of Classroom and Laboratory Conditions

Source	df	MS	F	p	η^2
Overall course grade					
Classroom	2	675.05	48.85	<.001	.462
Laboratory	1	1.36	0.10	.754	.001
Classroom × Laboratory	2	28.00	2.03	.137	.034
Error	114	13.82			
Recall and recognition score					
Classroom	2	39.44	15.09	<.001	.209
Laboratory	1	0.47	0.18	.673	.002
Classroom × Laboratory	2	0.33	0.12	.883	.002
Error	114	2.61			
Problem-solving score					
Classroom	2	24.08	13.55	<.001	.192
Laboratory	1	421.50	237.18	<.001	.675
Classroom × Laboratory	2	4.36	2.46	.090	.041
Error	114	1.78			
Course satisfaction ratings					
Classroom	2	1420.88	276.87	<.001	.829
Laboratory	1	401.50	78.24	<.001	.407
Classroom × Laboratory	2	4023.58	784.03	<.001	.932
Error	114	5.13			

▨ **Table 9.12.**

Table X

Two-Way (Classroom and Laboratory) Analyses of Variance for the Overall Grade, Recall and Recognition and Problem-Solving Scores, and Course Satisfaction Ratings

Variable and source	MS	F	p	η^2
Overall grade				
Classroom[a]	675.05	48.85	<.001	.462
Laboratory[b]	1.36	0.10	.754	.001
Classroom × Laboratory[a]	28.00	2.03	.137	.034
Recall and recognition score				
Classroom[a]	39.44	15.09	<.001	.209
Laboratory[b]	0.47	0.18	.673	.002
Classroom × Laboratory[a]	0.33	0.12	.883	.002
Problem-solving score				
Classroom[a]	24.08	13.55	<.001	.192
Laboratory[b]	421.50	237.18	<.001	.675
Classroom × Laboratory[a]	4.36	2.46	.090	.041
Course satisfaction				
Classroom[a]	1,420.88	276.87	<.001	.829
Laboratory[b]	401.50	78.24	<.001	.407
Classroom × Laboratory[a]	4,023.58	784.03	<.001	.932

[a]$df = 2, 144.$ [b]$df = 1, 144.$

> In this table, the degrees of freedom are in lettered table notes rather than in the body of the table.

Another way to present the same ANOVA results is shown in Table 9.13. In Table 9.13 the *F* ratios and significance levels are presented for the main and interaction effects for each ANOVA. Because the degrees of freedom are the same for each of the dependent variables presented, they can easily be listed in a single column. As shown, letters may be used as abbreviations for the independent variables.

Table 9.13.

Table X

Analysis of Variance Results for Four Course-Outcome Measures as a Function of Classroom and Laboratory Conditions

Source	df	Overall course grade	Recall and recognition score	Problem-solving score	Course satisfaction ratings
Classroom (C)	2	48.85*	15.09*	13.55*	276.87*
Laboratory (L)	1	0.10	0.18	237.18*	78.24*
C × L	2	2.03	0.12	2.46	784.03*
S/CL	114	(13.82)	(2.61)	(1.78)	(5.13)

Note. Values in parentheses represent mean square errors. S/CL = within-cell variance.

*p < .001.

> Mean squares for main and interaction effects and effect sizes are not included in this table.

> In this sample table, asterisk notes are used to express significance levels. Note, however, that APA Style recommends reporting exact significance levels.

Table 9.14.

Table X

> Identifying the number of participants included for each variable is particularly useful when the number of participants differs from variable to variable.

Two-Way (Classroom and Laboratory) Analyses of Variance for Overall Course Grade, Recall and Recognition and Problem-Solving Scores, and Course Satisfaction Ratings

Source	Overall course grade ($N = 120$)	Recall and recognition score ($N = 118$)	Problem-solving score ($N = 120$)	Course satisfaction ratings ($N = 120$)
Classroom (three levels)	48.85*	15.09*	13.55*	276.87*
Laboratory (two levels)	0.10	0.18	237.18*	78.24*
Classroom × Laboratory	2.03	0.12	2.46	784.03*
Within-cell variance	(13.82)	(2.61)	(1.78)	(5.13)

Note. Values in parentheses represent mean square errors.

*p < .001.

> Degrees of freedom have been omitted, and the numbers of participants for each analysis have been included.

> Asterisk notes are used to express significance levels. Note, however, that APA Style recommends reporting exact significance levels.

Example 9.5

In this example there are three independent variables. There are three levels of classroom instruction, two levels of laboratory instruction, and an additional independent variable: two levels of course content. The two levels of course content differ as to the specific material that is to be taught. Students are randomly assigned to either an introductory calculus course or an introductory psychology course. With the increase in the number of independent variables in the analyses, the table of means and standard deviations becomes more complex. The means and standard deviations of a $3 \times 2 \times 2$ design are presented in Table 9.15. The dependent variable is overall course grade.

Variables for Example 9.5

Independent Variables

1. Classroom instruction (lecture, textbook, or multimedia computer-assisted instruction)
2. Laboratory instruction (group discussion or problem solving)
3. Course (introductory calculus or introductory psychology)

Dependent Variable

1. Overall course grade

Sometimes the number of participants per cell is not equal. When this is the case, it may be helpful to include the number of participants per cell in the table of means and standard deviations. See Chapter 3 for examples.

An ANOVA table for more than two independent variables looks similar to a two-way ANOVA table, except that one additional main effect and three more interaction effects are included. The more independent variables there are, the more main effects and interaction effects there are to be presented (see Table 9.16). There are several ways to present ANOVA results for more than two independent variables. (See Tables 9.8 and 9.9 for alternate presentations.)

Table 9.15.

Table X

See Chapter 3 for other examples of format for tables of means and standard deviations.

Means and Standard Deviations for Class Conditions as a Function of Laboratory Condition and Course Content

Class	Group discussion		Problem solving	
	M	*SD*	*M*	*SD*
Introductory calculus				
Lecture	72.76	2.10	71.19	2.48
Textbook	70.56	1.80	72.31	3.08
Multimedia	77.76	6.12	79.35	4.95
Introductory psychology				
Lecture	72.50	2.03	71.18	1.81
Textbook	70.44	2.25	72.06	2.77
Multimedia	80.56	6.45	77.22	4.56

■ **Table 9.16.**

Table X

Analysis of Variance Results for Main Effects and Interaction Effects of Classroom, Laboratory, and Course Instruction on Grade

Variable	df	MS	F	p	η^2
Main effect of classroom (CL)	2	675.06	48.19	<.001	.472
Main effect of Laboratory (L)	1	1.36	0.10	.756	.001
Main effect of course (CO)	1	0.00	0.00	.993	.000
CL × L	2	28.00	2.00	.140	.036
CL × CO	2	0.82	0.06	.943	.001
L × CO	1	19.37	1.38	.242	.013
CL × L × CO	2	20.76	1.48	.232	.027
Within-cells error	108	14.01			

Within-Subjects, Mixed, and Hierarchical Designs: What Are They?

Within-subjects, mixed, and hierarchical designs are broad categories that include repeated measures designs, split-plot designs, nested designs, and numerous others that consist of variations of between-subjects and within-subjects designs.

What Tables Are Used?

Two types of tables usually are used: (a) a table that identifies the means and standard deviations for each cell of the design and (b) an ANOVA summary table. The ANOVA summary table could include the degrees of freedom, sums of squares, mean squares, *F* ratios, *p* values for the sources, and effect sizes, or some of this information could be excluded. (See the factorial ANOVA section for examples.) Also, depending on the complexity of the analysis, an ANOVA table may not be required; it may be possible to simply present the results within the text (this is more frequent in journal articles).

Example 9.6 is a mixed design, with one between-subjects variable and one within-subjects variable. The sample table presented (Table 9.18) can easily be adapted to more complex mixed designs, such as those with two between-subjects variables and one within-subjects variable (more main effects and interaction effects would have to be added in rows for the between-subjects and within-subjects sources of variance). For hierarchical designs, if the hierarchical design includes a within-groups variable, then the ANOVA table would look similar to Table 9.18 (with the between-subjects sources of variance, degrees of freedom, sums of squares, mean squares, *F* ratios, *p* values, and effect sizes shown first, followed by the within-subjects sources of variance). Other hierarchical designs should be organized such that the sources of variance (i.e., main effects, interaction effects, nested, and error sources) are organized in a logical manner. The table of the mixed design (Table 9.18) and the tables of factorial between-subjects designs presented earlier in this chapter can be used as guidelines.

"Play It Safe" Table

There are two "Play It Safe" tables in this section. Table 9.17 includes means and standard deviations, and Table 9.18 includes the degrees of freedom, sums of squares, mean squares, F ratios, p values, and effect sizes for the different sources. Tables 9.17 and 9.18 are the "Play It Safe" tables for a mixed design with one between-subjects and one within-subjects variable. These tables are typically found in theses and reports. However, in many journal articles authors will present only a table of means and present the F ratios, degrees of freedom, p values, and effect sizes in text form.

Example 9.6

Researchers wished to determine the long-term effects of two new anxiety-reducing drugs on individuals who experienced major work-related anxiety for 1 year. There are four levels of this between-subjects variable (i.e., four different medication conditions), with 10 people in each condition:

1. *Drug A*. A new anxiety-reducing drug.
2. *Drug B*. A second new anxiety-reducing drug.
3. *Drug C*. An anxiety-reducing drug currently being prescribed by most physicians.
4. *Drug D*. A placebo.

Each participant's anxiety was measured using a 20-item work-related anxiety questionnaire on several occasions. There are five levels of this within-subjects variable:

1. *Time 1*. Tested immediately after taking the medication.
2. *Time 2*. Tested 1 month after taking the medication.
3. *Time 3*. Tested 3 months after taking the medication.
4. *Time 4*. Tested 6 months after taking the medication.
5. *Time 5*. Tested 1 year after taking the medication.

Thus, there are two independent variables (medication and testing time). The dependent variable is anxiety as measured using a 20-item work-related anxiety questionnaire.

Variables for Example 9.6

Independent Variables

1. Medication (Drug A, B, C, or D)
2. Testing time (Times 1, 2, 3, 4, and 5)

Dependent Variable

1. Anxiety

Table 9.17

Table X

Means and Standard Deviations for Four Drugs and Five Testing Times

Testing time	Drug A	Drug B	Drug C	Drug D
Time 1				
M	39.0	66.8	67.9	40.0
SD	6.0	3.70	4.41	6.78
Time 2				
M	46.7	63.8	66.3	49.6
SD	7.36	4.47	7.22	6.17
Time 3				
M	50.7	58.9	65.2	66.3
SD	6.00	6.82	7.22	5.29
Time 4				
M	52.1	57.6	65.2	50.8
SD	6.26	8.55	8.18	5.37
Time 5				
M	55.6	53.5	65.3	41.3
SD	7.35	8.76	5.85	5.25

See Chapter 3 for other examples of format for tables of means and standard deviations.

Table 9.18

Table X

Analysis of Variance Results for Medication and Time Variables

Source	df	SS	MS	F	p	η^2
		Between subjects				
Drug	3	10,450.98	3,483.66	26.5	<.001	.69
Error 1	36	4,732.44	131.46			
		Within subjects				
Time	4	1,186.72	296.68	14.97	<.001	.29
Drug × Time	12	5,991.92	499.33	25.19	<.001	.68
Error 2	144	2,854.56	19.82			

This table, along with a table of means and standard deviations, is the "Play-It-Safe" table for a mixed design with one between-subjects and one within-subjects variable. Note that for journals, the *F* statistic along with the degrees of freedom, *p* value, and effect size are frequently presented in the text and not in a table. A separate column for *p* values generally is used only when they are not the same for all results; the column is shown here for illustrative purposes only.

10

Multivariate Analysis of Variance

What Is It?

The multivariate analysis of variance (MANOVA) is an extension of the analysis of variance (ANOVA) to the situation in which there is more than one continuous dependent variable. As in ANOVA, the analysis allows a researcher to identify whether different levels of the categorical independent variable or variables have effects on the dependent variables. In addition, MANOVA examines the associations among dependent variables.

What Tables Are Used?

There are three tables that are especially relevant for a MANOVA. The three most commonly used tables are (a) a table of means and standard deviations for the dependent variables (Table 10.1), (b) a table of correlations among the dependent variables (Table 10.2), and (c) a multivariate and univariate ANOVA summary table (Tables 10.3 and 10.4). The results reported depend on the nature of the data as well as the relevance of the tables to the hypotheses and the context in which the tables are being presented (e.g., journal article vs. thesis).

"Play It Safe" Table

The "Play It Safe" tables are Tables 10.1, 10.2, and 10.3. To be comprehensive, all three tables should be included.

Example 10.1

The example used in this chapter is a reading study. The researchers were interested in sex differences in reading impairment. Boys and girls with and without dyslexia were tested on four measures of reading ability. Thus, this is a 2 (boys or girls) × 2 (with dyslexia or without dyslexia) MANOVA with four dependent variables. The dependent variables are four measures of reading ability: the Famous Authors Test, a verbal IQ test, the Ingersoll Reading Test, and average speed on a nonword pronunciation task (nonword performance).

Variables for Example 10.1

Independent Variables

1. Gender (boys vs. girls)
2. Reading impairment (with dyslexia vs. without dyslexia)

Dependent Variables

1. Famous Authors Test
2. Verbal IQ
3. Ingersoll Reading Test
4. Nonword pronunciation speed (nonword performance)

Table 10.2, an example of a correlation table, is especially relevant if there are significant correlations among some or all of the dependent variables. In the reading study example in this chapter, the four dependent variables are related because they all measure aspects of reading ability. Thus, some of these variables are correlated. The results of the multivariate and univariate ANOVAs are presented together in Table 10.3, whereas Table 10.4 illustrates a somewhat different version of a MANOVA summary table.

Table 10.1.

See Chapter 3 for other examples of format for tables of means and standard deviations.

Table X

Mean Scores and Standard Deviations for Measures of Reading Ability as a Function of Gender and Reading Impairment

Group	Famous Authors Test		Verbal IQ		Ingersoll Reading Test		Nonword performance	
	M	*SD*	*M*	*SD*	*M*	*SD*	*M*	*SD*
Boys								
With dyslexia	26	3.2	105	8.2	15	3.2	1,081	103.1
Without dyslexia	28	5.3	106	11.6	24	3.3	830	64.3
Girls								
With dyslexia	25	4.5	98	18.1	16	4.1	1,018	77.8
Without dyslexia	30	6.6	107	7.0	26	4.9	815	60.5

Note. Nonword performance = average response latency (in milliseconds) on a nonword pronunciation task.

Table 10.2.

See Chapter 7 for other examples of format for correlation tables.

Table X

Correlation Coefficients for Relations Between Four Measures of Reading Ability

Measure	1	2	3	4
1. Famous Authors Test	—			
2. Verbal IQ	.03	—		
3. Ingersoll Reading Test	.18*	.41**	—	
4. Nonword performance	−.22*	−.16	−.64***	—

Note. Nonword performance = average response latency (in milliseconds) on a nonword pronunciation task.
*$p < .05$. **$p < .01$. ***$p < .001$.

Table 10.3.

This table presents the results of the multivariate and univariate analyses of variance together.

Table X

Multivariate and Univariate Analyses of Variance for Reading Measures

Source	Multivariate			Univariate											
				Famous Authors Test			Verbal IQ			Ingersoll Reading Test			Nonword performance		
	F^a	p	η^2	F^b	p	η^2	F^b	p	η^2	F^b	p	η^2	F^b	p	η^2
Gender (G)	3.11	.018	.10	0.42	.520	.00	1.45	.230	.01	3.03	.084	.03	7.48	.007	.06
Reading impairment (R)	118.44	<.001	.81	14.45	<.001	.11	5.93	.016	.05	174.33	<.001	.60	251.71	<.001	.68
G × R	2.43	.052	.08	4.07	.046	.03	2.80	.097	.02	0.01	.908	.00	2.79	.098	.02

Note. Multivariate *F* ratios were generated from Pillai's statistic. Nonword performance = average response latency (in milliseconds) on a nonword pronunciation task. [a]Multivariate *df* = 4, 113. [b]Univariate *df* = 1, 116.

Table 10.3, along with Tables 10.1 and 10.2 (means and standard deviations and correlations, respectively), is the "Play It Safe" table for a MANOVA.

Table 10.4.

This is a slightly different version of a MANOVA summary table than Table 10.3.

Table X

Multivariate and Univariate Analyses of Variance F Ratios for Gender × Reading Impairment Effects for Reading Measures

		ANOVA $F(1, 116)$			
Variable	MANOVA $F(4, 113)$	Famous Authors Test	Verbal IQ	Ingersoll Reading Test	Nonword performance
Gender (G)	3.11*	0.42	1.45	3.03	7.48**
Reading impairment (R)	118.44***	14.45***	5.93*	174.33***	251.71***
G × R	2.43	4.07*	2.80	0.01	2.79

Note. *F* ratios are Wilks's approximation of *F*. ANOVA = univariate analysis of variance; MANOVA = multivariate analysis of variance; Nonword performance = average response latency (in milliseconds) on a nonword pronunciation task.
*$p < .05$. **$p < .01$. ***$p < .001$.

Analysis of Covariance

What Is It?

The analysis of covariance (ANCOVA) is an extension of the analysis of variance (ANOVA) and is used when the effects of a covariate, or uncontrolled source of variation, need to be removed from the ANOVA. An ANCOVA is used when there is one dependent variable.

What Tables Are Used?

There are two tables that are particularly relevant when presenting data analyzed with an ANCOVA: (a) a table of means and standard deviations for the dependent variable (often posttest scores) and the covariate (often pretest scores) as a function of the independent variable or variables (Table 11.1) and (b) the ANCOVA summary table (Table 11.2 or 11.3). The table of means and standard deviations and the ANCOVA summary table sometimes are combined (Table 11.4). For ANCOVA results researchers typically present adjusted means, but depending on the research questions may present unadjusted means, and examples of both are presented (unadjusted means in Table 11.1, adjusted means in Table 11.4).

"Play It Safe" Table

The "Play It Safe" choice for ANCOVA tables is Table 11.1 (means and standard deviations) with Table 11.2 (ANCOVA summary table).

Example 11.1

In an educational study, a researcher developed a new aid for teaching seventh-grade students about electric circuits. The aid was a transparent circuit board that the students could experiment with on their own. The researcher wanted to know whether the students would learn more using the aid if they (a) explored it individually without any instruction, (b) were given written instructions about it, or (c) watched a demonstration of how it works. The researcher also wanted to know whether the students would learn more if an eighth-grade student tutor was assigned to help them. The researcher wanted to control for the differences among the students in terms of the amount they knew about electric circuits before they began the session with the circuit board. This was measured with a presession written test (*pretest*). After the session with the circuit board, the students' learning was assessed with a written test (*posttest*). Thus, the independent variables are instruction condition and whether a tutor assisted students during the session. The covariate is the students' presession (pretest) knowledge. The dependent variable is the students' postsession (posttest) knowledge.

Variables for Example 11.1

Independent Variables

1. Instruction condition (no instruction, written instruction, or demonstration)
2. Tutor help (presence vs. absence of eighth-grade student tutor)
3. Pretest score (covariate)

Dependent Variable

1. Posttest score

Table 11.1.

See Chapter 3 for other examples of format for tables of means and standard deviations.

Table X

Pre- and Posttest Mean Scores and Standard Deviations as a Function of Instruction Condition and Tutor Help

	Pretest		Posttest	
Source	M	SD	M	SD
No instruction				
Tutor help	48.80	11.82	73.00	8.08
No tutor help	49.10	9.12	72.40	7.79
Written instruction				
Tutor help	55.50	10.70	78.20	9.74
No tutor help	62.80	8.15	67.30	9.98
Demonstration				
Tutor help	70.40	6.72	83.30	9.73
No tutor help	68.60	11.76	73.70	7.85

Table 11.2.

This table, along with Table 11.1 for means and standard deviations, is the "Play It Safe" table for ANCOVA results.

Table X

Analysis of Covariance of Posttest Knowledge Scores as a Function of Instruction Condition and Tutor Help, With Pretest Knowledge Scores as Covariate

Source	df	SS	MS	F	p	η^2
Covariate	1	335.28	335.28	4.49	.039	.078
Instruction condition (IC)	2	778.02	389.01	5.21	.009	.164
Tutor help (TH)	1	636.08	636.08	8.52	.005	.139
IC × TH	2	276.96	136.98	1.84	.170	.065
Error	53	3,955.02	74.62			
Total	59	5,971.65				

■ **Table 11.3**

Table X

Analysis of Covariance for Instruction Condition and Tutor Help, With Pretest Knowledge Scores as Covariate

Source	df	MS	F	p	η^2
Pretest knowledge (covariate)	1	335.28	4.49	.039	.078
Instruction condition (IC)	2	389.01	5.21	.009	.164
Tutor help (TH)	1	636.08	8.52	.005	.139
IC × TH	2	136.98	1.84	.170	.065
Error	53	74.62			

Example 11.2

The researcher conducted a second experiment to determine how different types of cognitive skills changed as a result of exposure to the circuit board. In this experiment, a new group of 60 students participated in the circuit board session. Half of the students received tutor help, and half did not. Before the session with the circuit board, the students were tested for five cognitive skills: general problem solving, science problem solving, electronics problem solving, creativity, and spatial rotation (pretest). They were tested for these skills again following the session (posttest). The researcher performed an ANCOVA to determine how each of these skills was influenced by participation in the session while controlling for presession skill levels. The independent variable is tutor help, the covariate is presession cognitive skills (pretest), and the dependent variable is postsession cognitive skills (posttest).

Variables for Example 11.2

Independent Variables

1. Tutor help (presence vs. absence of eighth-grade student tutor)
2. Presession (pretest) cognitive skills (covariate; general problem solving, science problem solving, electronics problem solving, creativity, and spatial rotation)

Dependent Variable

1. Postsession (posttest) cognitive skills (general problem solving, science problem solving, electronics problem solving, creativity, and spatial rotation)

Table 11.4.

A full analysis of covariance (ANCOVA) summary table often is not necessary, as illustrated in the following table, which is a combination of a table of means and standard deviations and an ANCOVA summary table.

Table X

Pretest Means, Adjusted Posttest Means, Standard Deviations, and Analysis of Covariance Results for Five Cognitive Skills

Cognitive skill	Tutor help				No tutor help				$F(1, 57)$	p	η^2
	Pretest		Posttest		Pretest		Posttest				
	M	SD	M	SD	M	SD	M	SD			
GPS	40.17	10.70	42.96	8.93	40.05	8.85	36.41	10.04	7.16	.010	.112
SPS	31.16	10.56	31.45	8.40	30.56	10.36	29.70	8.75	0.62	.434	.011
EPS	20.21	6.29	27.99	7.97	18.87	6.63	22.24	6.21	9.26	.004	.140
Creativity	12.72	3.62	20.63	6.05	11.27	2.71	12.97	4.41	30.12	<.001	.346
Spatial rotation	22.65	5.20	22.07	6.01	18.30	6.70	17.91	6.23	6.13	.016	.097

Note. GPS = general problem solving; SPS = science problem solving; EPS = electronics problem solving.

12

Multivariate Analysis of Covariance

What Is It?

The multivariate analysis of covariance (MANCOVA) is the extension of an analysis of covariance (ANCOVA) to the situation in which there are several dependent variables and one or more covariates. As such, MANCOVA tests whether the independent variables have an effect on the dependent variables after variance accounted for by the covariate or covariates has been removed.

What Tables Are Used?

There are three types of tables that are especially relevant when presenting data that have been analyzed using MANCOVA: (a) a table of means or adjusted means and standard deviations as a function of the independent variables (see Chapter 11 on ANCOVA), (b) a table of correlations among the dependent variables (see Chapter 10 on multivariate analysis of variance [MANOVA]), and (c) the summary table of univariate analyses of variance (see Chapter 9) and possibly MANOVAs (see Chapter 10). However, researchers do not always include all three of these tables when presenting the results of a MANCOVA. Sometimes the multivariate results are presented only in the text. Additionally, in some cases, the tables described in (a) and (b) or (a) and (c) are combined into a single table.

"Play It Safe" Table

The "safe" choice is to present each of the three types of tables described.

13

Cluster Analysis

What Is It?

Cluster analysis refers to a variety of techniques used to determine the underlying structure or natural grouping of a set of entities by illustrating which of those entities are most closely related on the basis of a set of descriptors (e.g., attitudes, interests, symptoms, or traits). The underlying structure or natural groupings often are referred to as *clusters*. There are various methods (e.g., agglomerative hierarchical clustering and divisive hierarchical clustering) and measures (e.g., square Euclidean distance and Pearson product–moment correlation) for calculating distances between descriptors, and there are various methods (e.g., between-groups linkage, nearest neighbor, and farthest neighbor) for combining descriptors into clusters. Agglomerative hierarchical clustering is illustrated in this chapter. The squared Euclidean distance is the measure of distance, and the between-groups linkage is the method used to determine whether two entities should be combined.

What Tables Are Used?

The results of cluster analyses are never presented in a table. The various clusters generally are presented in either a dendrogram or a figure. If the researcher has numerous descriptors and wishes to determine whether a combination of these descriptors forms different clusters, a dendrogram is used. If the researcher wishes to determine whether profiles can be created for individuals (or objects or animals) based on a set of variables, then a figure is used. A dendrogram is illustrated in Example 13.1 (Figure 13.1), and a figure is presented in Example 13.2 (Figure 13.2). For other examples of figures, see the sister volume to this one, *Displaying Your Findings*.[1]

---

[1]Nicol, A. A. M., & Pexman, P. M. (2010). *Displaying your findings: A practical guide for creating figures, posters, and presentations* (6th ed.). Washington, DC: American Psychological Association.

In addition to dendrograms or figures, a table of means and standard deviations for all of the variables should be included (see Table 13.1). If one's purpose is to examine differences between profiles of individuals on the variables that make up those profiles, then a table of means and standard deviations and the results of any tests of group differences that may have been conducted (e.g., *F* tests, *t* tests, or post hoc analyses) are also presented (see Table 13.2).

"Play It Safe" Table

The most comprehensive table is one including the means and standard deviations and the results of any tests of group differences conducted between the clusters (see Table 13.2). In addition, a dendrogram or figure should be included to illustrate the results of the cluster analysis (see Figures 13.1 and 13.2, respectively).

Example 13.1

Two researchers videotaped 192 individuals. Each videotape consisted of a person having a conversation with a close friend about what constitutes a liberal education. The researchers coded numerous behaviors such as irrelevant hand gestures, smiling, eye contact, and so on. A total of 10 behaviors were coded. The researchers wished to determine whether an individual's interpersonal communication style could be described by clusters of these 10 behaviors.

Variables for Example 13.1

1. Establishes eye contact
2. Listens attentively
3. Relates topic to other person
4. Interrupts frequently
5. Relates topic only to self
6. Asks questions
7. Does not answer questions
8. Uses irrelevant hand gestures
9. Varies speed of speech
10. Smiles

■ **Figure 13.1.**

According to APA Style, figures are presented after tables in a manuscript. Each figure is presented on a separate page and with a descriptive figure caption. Both figure and figure caption should appear on the same page.

This is a "Play It Safe" figure for the results of a cluster analysis.

A different scale may be used depending on the analyses that were conducted.

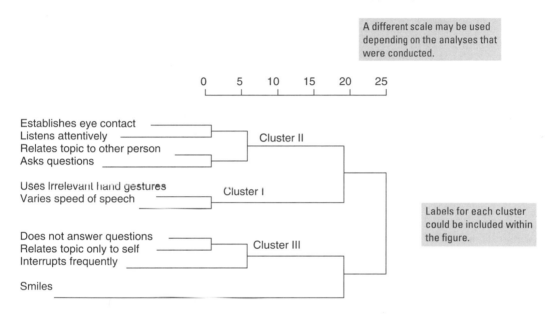

Labels for each cluster could be included within the figure.

Figure X. Dendrogram based on the results of the agglomerative cluster analysis.

■ **Table 13.1.**

Table X

This table presents the means and standard deviations for each variable. For more example formats of tables of means and standard deviations, see Chapter 3.

Descriptive Statistics for 10 Communication Behaviors (N = 192)

Communication behaviors	M	SD
Establishes eye contact	28.96	21.59
Listens attentively	29.77	19.17
Relates topic to other person	27.39	24.46
Interrupts frequently	31.33	19.87
Relates topic only to self	30.39	21.80
Asks questions	30.22	21.44
Does not answer questions	30.04	22.22
Uses irrelevant hand gestures	29.93	21.61
Varies speed of speech	30.87	21.90
Smiles	36.13	23.58

This is a "Play It Safe" table for the results of a cluster analysis.

Example 13.2

Two researchers wished to determine whether profiles exist for various communication styles. They videotaped 88 participants having a conversation with a close friend about the physiological, social, and economic benefits of being a vegetarian. The researchers obtained three measures of communication: one physical, one emotional, and one conversational. They wished to determine whether communication profiles could be formed using these three measures and, if so, what form these profiles would take.

Variables for Example 13.2

1. Physical
2. Emotional
3. Conversational

▦ Figure 13.2.

This is a "Play It Safe" figure for the results of a cluster analysis. Table 13.2 is also required to present the differences among the various profiles.

For more information regarding figures, see the sixth editions of the *Publication Manual* and *Displaying Your Findings: A Practical Guide for Creating Figures, Posters, and Presentations.*

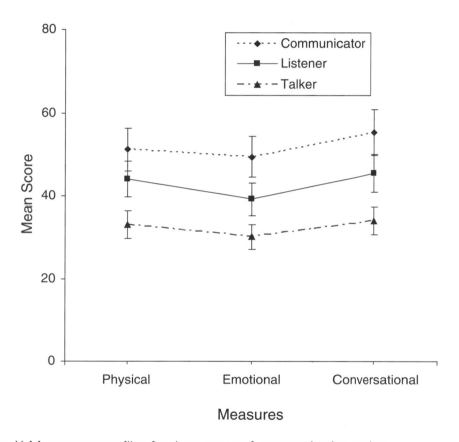

Figure X. Mean-score profiles for three types of communication styles.

Table 13.2.

Table X

Between-Groups Differences for Physical, Emotional, and Conversational Measures

Measure	Group 1 (communicators; $n = 18$)		Group 2 (listeners; $n = 16$)		Group 3 (talkers; $n = 54$)		$F(2, 85)$	p	η^2
	M	SD	M	SD	M	SD			
Physical	51.22	6.10	44.09	5.16	33.12	11.04	28.18	<.001	.40
Emotional	49.44	8.44	39.20	5.22	30.10	14.01	18.69	<.001	.31
Conversational	55.28	10.86	45.50	5.77	34.00	11.36	30.08	<.001	.41

If different analyses were conducted to determine group differences, they would be presented in the column in which *F* ratios are located. Alternatively, some researchers prefer to place this information in the Results section of the text rather than in a table. A separate column for *p* values generally is used only when they are not the same for all results; the column is shown here for illustrative purposes only.

For more examples of how to present *F* ratios, the results of *t* tests, and the results of post hoc analyses, see Chapters 9, 5, and 6, respectively.

Log-Linear Analysis

14

What Is It?

Log-linear analysis is used to examine the relation between two or more categorical variables to determine the best model (it tests main and interaction effects) that will account for the observed frequencies. In log-linear analysis, all of the categorical variables are considered to be independent variables. There are several ways to conduct log-linear analysis (e.g., hierarchical or nonhierarchical), but each requires that results be presented in a similar manner. A hierarchical log-linear analysis using backward elimination is used as the example in this chapter.

What Tables Are Used?

There are three tables that are used to present the results of log-linear analysis: (a) a table of the observed frequencies for the variables (see Tables 14.1 and 14.4), (b) a table of the log-linear parameters and the goodness-of-fit tests (see Table 14.2), and (c) a table of the goodness-of-fit index for each step in the analysis (see Table 14.3). Table 14.2 is not necessary if the final model consists of only a few main effects. In that case, the results can be reported in the text of the Results section rather than in a table.

"Play It Safe" Table

The "safe" choices for log-linear analysis include Table 14.1 (observed frequencies for all of the variables), Table 14.2 (log-linear parameters, z values, and goodness-of-fit tests) or Table 14.3 (goodness-of-fit index for each step), and Table 14.4 (observed frequencies for significant interactions). If many variables are included in the analysis, Table 14.1 can be omitted, because a table of observed frequencies for all of the variables would be too complex.

Example 14.1

A physical education instructor wished to determine the relations among gender, sport, and major. A total of 503 college students completed the survey. The instructor categorized sport activities into three categories: (a) individual sport (e.g., jogging); (b) team sport (e.g., hockey); and (c) a combination of individual sport and team sport, which was labeled *both* (e.g., swim team). The instructor grouped the students' course majors into two broad categories: arts or science. Thus, there are three variables in this analysis: gender, sport, and major.

The observed cell frequencies are presented in Table 14.1. The results of the log-linear analysis conducted on the data are presented in Tables 14.2 and 14.3. The observed cell frequencies for the interaction included in the final model are given in Table 14.4.

Variables for Example 14.1

1. Gender (men or women)
2. Sport (individual, team, or both)
3. Major (arts or science)

■ **Table 14.1.**

Table X

Observed Frequencies and Percentages for Sport, Gender, and Major

Sport	Women	Men
	Arts major	
Team	64 (13)	47 (9)
Individual	30 (6)	14 (3)
Both	14 (3)	14 (3)
	Science major	
Team	30 (6)	14 (3)
Individual	91 (18)	45 (9)
Both	93 (18)	47 (9)

Note. Percentages appear in parentheses.

This table is not always used, particularly when there are many variables included in the log-linear analysis, which would make the table difficult to read. In such a case, the researcher may wish to present only the observed frequencies for effects that are included in the model (e.g., significant interactions).

This table, along with Table 14.2 or 14.3 and Table 14.4, is the "Play It Safe" table for presenting the results of log-linear analysis.

■ **Table 14.2.**

Table X

Log-Linear Parameter Estimates, Values, and Goodness-of-Fit Index for Gender, Sport, and Major

Effect	λ	z	p
Sport × Major	.764	9.90	<.001
	−.281	−3.66	<.001
Gender	.264	4.76	<.001
Sport	.005	0.07	.944
	.082	1.07	.284
Major	−.280	−5.04	<.001

Note. $G^2(5, N = 503) = 5.73, p = .333$.

Note that only the effects that describe the final model, as determined by the log-linear analysis, are included.

G^2 is the likelihood ratio chi-square with 5 degrees of freedom. λ is lambda and is the parameter estimate or effect coefficient.

Standard errors could be presented with z values.

■ **Table 14.3.**

The researcher may wish to present the goodness-of-fit index for each step of the analysis as shown in this table.

Table X

Summary of Hierarchical Deletion Steps Involved in Arriving at Final Model

Step	Model	df	G^2	p	Term deleted	Δdf	ΔG^2	Δp
1	(G × S)(G × M)(S × M)	2	1.93	.381	G × S	2	0.70	.705
2	(G × M)(S × M)	4	2.63	.621	G × M	1	3.10	.078
3	(S × M)(G)	5	5.73	.333	G	1	40.06	<.001

Note. G = gender; S = sport; M = major.

Note that the three-way interaction (Gender × Sport × Major) is not presented in this table because it is the saturated model and therefore has 0 degrees of freedom and a G^2 also equal to 0. Also, terms listed imply that lower order terms (i.e., main effects in this example) are included. The last step presented is the one in which Δp is significant.

■ **Table 14.4.**

A frequency table is useful for displaying any interactions that may be in the final model.

Table X

Cross-Tabulation of Observed Frequencies and Percentages for Sport × Major Interaction

	Major	
Sport	Arts	Science
Team	111 (22)	44 (9)
Individual	44 (9)	136 (27)
Both	28 (6)	140 (28)

Note. Percentages appear in parentheses.

15

Discriminant Function Analysis

What Is It?

Discriminant function analysis (also known as *discriminant analysis*) allows prediction of group membership (when groups are different levels of the categorical dependent variable) from a set of predictor variables. That is, the analysis determines which predictor variables discriminate between two or more groups.

What Tables Are Used?

If the analysis is a stepwise discriminant function analysis, the four tables most commonly used to present results are (a) a table of means and standard deviations for predictor variables as a function of group (Table 15.1), (b) a table of discriminant function results with Wilks's lambda results for each step (Table 15.2), (c) a table of discriminant function coefficients (Table 15.3 or 15.4), and (d) a table of results of classification analysis for the discriminant function analysis (Table 15.5 or 15.6). Occasionally, there is also a centroids plot if there are three or more groups (levels of the dependent variable) involved in the analysis (Figure 15.1).

If the predictor variables in the discriminant function analysis were entered directly (and not stepwise), then the tables used are the same with one exception: The table for Wilks's lambda results usually is omitted, and the Wilks's lambda results are presented within the text instead of in a table.

"Play It Safe" Table

The "Play It Safe" tables for a stepwise discriminant function analysis are Tables 15.1, 15.2, 15.3, and 15.5. The "Play It Safe" tables for other discriminant function analyses are Tables 15.1, 15.3, and 15.5. The "safe" choice for stepwise and all other discriminant function analyses also includes a centroids plot (Figure 15.1), particularly if there are three or more groups (levels of the dependent variable).

Example 15.1

In this study, the researchers were interested in the factors that can predict chess skill (the dependent variable). They recruited three groups of participants: chess experts, chess amateurs, and chess novices. There were 30 participants in each group. The researchers measured participants' performance on several predictor variables to determine whether performance on those variables could predict group membership. The predictor (independent) variables are spatial ability, problem-solving ability, map-reading skill, and accuracy of visual imagery.

Variables for Example 15.1

Independent Variables

1. Spatial ability
2. Problem-solving ability
3. Map-reading skill
4. Accuracy of visual imagery

Dependent Variable

1. Chess skill

▦ Table 15.1.

This is a "Play It Safe" table for all types of discriminant function analysis.

For other examples of tables of means and standard deviations and tables of a priori contrasts, see Chapters 3 and 6, respectively.

Table X

Means and Standard Deviations of Predictor Variables as a Function of Chess Skill Group

	Experts		Amateurs		Novices	
Predictor variable	*M*	*SD*	*M*	*SD*	*M*	*SD*
Spatial ability	16.79$_{a,b}$	3.51	12.09$_{a,c}$	2.94	8.16$_{b,c}$	2.09
Problem-solving ability	89.14$_{d,e}$	4.29	73.07$_{d,f}$	3.95	66.57$_{e,f}$	5.94
Map-reading skill	15.70$_g$	4.39	12.30$_g$	3.54	13.87	4.10
Accuracy of visual imagery	37.29$_h$	1.85	34.87	3.93	33.33$_h$	4.87

Note. Means with the same subscript differ significantly at *p* < .01.

▦ Table 15.2.

This is a "Play It Safe" table for a stepwise discriminant function analysis. Information for other discriminant function analyses could be presented in the text.

Table X

Predictor Variables in Stepwise Discriminant Function Analysis

Step	Predictor variable	Variables in discriminant function	Wilks's λ	Equivalent *F*(2, 87)	*p*
1	Spatial ability	1	.397	66.20	<.001
2	Problem-solving ability	2	.199	175.20	<.001
3	Map-reading skill	3	.890	5.38	.006
4	Accuracy of visual imagery	4	.884	5.74	.005

■ **Table 15.3.**

This is a "Play It Safe" table for all types of discriminant function analyses.

Table X

Correlation of Predictor Variables With Discriminant Functions (Function Structure Matrix) and Standardized Discriminant Function Coefficients

Predictor variable	Correlation with discriminant functions		Standardized discriminant function coefficients	
	Function 1	Function 2	Function 1	Function 2
Spatial ability	.883	.213	.490	−.550
Problem-solving ability	.159	−.088	.760	.200
Map-reading skill	.109	.780	.198	.828
Accuracy of visual imagery	.537	−.595	.279	.182

■ **Table 15.4.**

This table illustrates that sometimes a table of coefficients presents only the structure matrix and that sometimes coefficients are presented to only two decimal places.

Table X

Correlations Between Discriminating Variables and Discriminant Functions (Function Structure Matrix)

Variable	Function 1	Function 2
Spatial ability	.88	.21
Problem-solving ability	.16	−.09
Map-reading skill	.11	.78
Accuracy of visual imagery	.54	−.59

Table 15.5.

This is a "Play It Safe" table for all types of discriminant function analyses.

Table X

Classification Analysis for Chess Skill

| | | Predicted group membership | | | | | |
| | | Experts | | Amateurs | | Novices | |
Actual group membership	*n*	*n*	%	*n*	%	*n*	%
Experts	30	30	100.0	0	0.0	0	0.0
Amateurs	30	1	3.3	23	76.7	6	20.0
Novices	30	0	0.0	5	16.7	25	83.3

Note. Overall percentage of correctly classified cases = 86.7%.

Table 15.6.

Results of the classification analysis could also be presented in this format.

Table X

Classification Analysis for Chess Skill

| | | Predicted group membership | | |
Actual group membership	*n*	Experts	Amateurs	Novices
Experts	30			
n		30	0	0
%		100.0	0.0	0.0
Amateurs	30			
n		1	23	6
%		3.3	76.7	20.0
Novices	30			
n		0	5	25
%		0	16.7	83.3

Note. Overall percentage of correctly classified cases = 86.7%.

■ **Figure 15.1.**

According to APA Style, figures are presented after tables in a manuscript. Each figure starts on a separate page and includes a descriptive figure caption. Both figure and figure caption are included on the same page, with the caption appearing below the figure.

The "safe" choice for stepwise and all other discriminant function analyses includes a centroids plot.

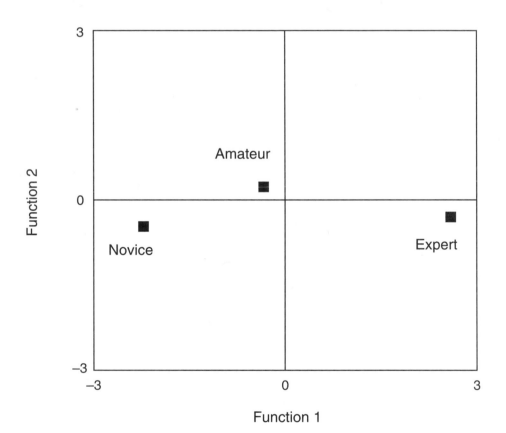

Figure X. Group centroids plot from discriminant function analysis.

16

Factor Analysis

What Is It?

Factor analysis is a multivariate statistical technique used to reduce the number of observed variables to a smaller number of latent variables, identified as *factors*. Different factor analytic procedures exist. The single term *factor analysis* is used in this chapter to refer to all types of factor analytic procedures.

What Tables Are Used?

Generally, one type of table is used that includes the factor loadings (rotated or unrotated) for the factors of interest. There are many ways these tables can be presented. For example, only the factor loadings may be presented (see Tables 16.4 and 16.11). Additional information regarding the factor analysis results can be included, such as the percentage of variance (total, common, or both), eigenvalues, and communalities (see Tables 16.1, 16.2, 16.3, and 16.10). Sometimes, information such as means, standard deviations, factor scores, and factor correlations is presented along with the factor loadings (see Tables 16.5, 16.6, 16.7, 16.8, and 16.9).

Example 16.1 presents the results of a single factor analysis without a factor rotation. Examples 16.2 and 16.3 present the results of a varimax rotation and an oblimin rotation, respectively. Example 16.4 presents the results of two separate factor analyses.

"Play It Safe" Table

The "Play It Safe" table would include (a) all of the factor loadings, eigenvalues, communalities, and percentages of variance for an unrotated factor analysis solution (see Table 16.1); (b) all of the factor loadings and communalities for an orthogonally rotated factor analysis solution (see Table 16.3; although for journal articles it is typical to find only the factor loadings); and (c) all of the factor loadings for an obliquely rotated solution (see Table 16.8).

For theses, a more thorough description of the factor analysis may be required. This could include a scree plot, which is not illustrated here, but an example can be found in *Displaying Your Findings*. Additionally, should a factor analysis of a measure be conducted, then it would be best to include descriptions of the items (as in Tables 16.3, 16.4, 16.5, and 16.8). If this is not possible, then a description of the items can be included in an appendix. Table 16.10 is an example of the "Play It Safe" table for the results of two separate factor analyses presented in a single table.

Example 16.1

Elementary school teachers ($N = 81$) were asked to rate the importance of several reasons for elementary school children failing their grade. A newly developed questionnaire, the 15-item Student Failure Questionnaire (SFQ), was provided to all participants. A principal-components factor analysis was conducted to identify the underlying factors of this new questionnaire.

The study consists of 15 variables: Each of the 15 items of the SFQ constitutes one variable, with each item identifying a potential reason for elementary school children failing their grade. Items labeled "(R)" are reverse scored.

Variables for Example 16.1

Student Failure Questionnaire Items

1. Student is excessively absent from school.
2. Student does extra work on his or her own. (R)
3. Education is taught in a second language.
4. Student asks questions about material. (R)
5. Student lacks adequate diet.
6. Student lacks support from school.
7. Student does not do homework.
8. Student lacks support from teacher.
9. Student has attention deficit disorder.
10. Student lacks support from other students.
11. Student appears to be of superior intellect. (R)
12. Student lacks parental support.
13. Student makes extra effort to participate in class. (R)
14. Student has learning disability.
15. Student enjoys doing schoolwork. (R)

Table 16.1.

This table presents the results of a factor analysis without any factor rotation.

This is the "Play It Safe" table for an unrotated factor analysis solution.

Item descriptions may be presented in an appendix or in a word table; see Chapter 22.

Table X

Factor Loadings From Principal Component Factor Analysis: Communalities, Eigenvalues, and Percentages of Variance for Items of the Student Failure Questionnaire

Item	Factor loading			Communality
	1	2	3	
1	.73	.45	−.46	.95
2	.66	.36	−.53	.85
3	.90	.18	.32	.94
4	.75	−.56	.25	.94
5	.29	−.43	−.34	.38
6	.86	.34	.25	.92
7	.75	.21	−.42	.78
8	.61	.67	−.17	.85
9	.54	−.36	.42	.60
10	.57	.61	−.17	.72
11	.76	−.46	.33	.90
12	.78	.21	−.29	.74
13	.33	−.43	.41	.46
14	.84	−.35	.15	.85
15	.79	−.14	.31	.74
Eigenvalue	7.34	2.56	1.72	
% of variance	48.93	17.07	11.47	

■ **Table 16.2.**

Table X

Eigenvalues, Percentages of Variance, and Cumulative Percentages for Factors of the 15-Item Student Failure Questionnaire

Factor	Eigenvalue	% of variance	Cumulative %
1	7.34	48.93	48.93
2	2.56	17.07	66.00
3	1.72	11.47	77.47

Occasionally, as shown by this table, a summary table of the eigenvalues, percentages of total variance, and cumulative percentages is used.

Example 16.2

In this example, the 15-item SFQ, described in Example 16.1, is subjected to a varimax orthogonal rotation.

■ **Table 16.3.**

Table X

*Summary of Items and Factor Loadings for Varimax Orthogonal Three-Factor Solution for
the Student Failure Questionnaire (N = 81)*

| | Factor loading | | | |
Item	1	2	3	Communality
3. Education is taught in a second language.	**.89**	.02	−.12	.80
14. Student has learning disability.	**.86**	−.02	.02	.74
9. Student has attention deficit disorder.	**−.86**	.11	.13	.76
5. Student lacks adequate diet.	**.79**	.04	−.14	.66
11. Student appears to be of superior intellect.	**.79**	.12	.03	.64
1. Student is excessively absent from school.	**.77**	−.16	.08	.62
7. Student does not do homework.	.05	**.86**	−.08	.76
15. Student enjoys doing schoolwork.	.01	**.79**	.10	.63
2. Student does extra work on his or her own.	.03	**.77**	.09	.60
13. Student makes extra effort to participate in class.	.10	**.75**	.32	.67
4. Student asks questions about material.	−.09	**.69**	.32	.59
12. Student lacks parental support.	.09	.09	**.90**	.82
6. Student lacks support from school.	.09	.11	**.89**	.81
8. Student lacks support from teacher.	.08	.06	**.75**	.58
10. Student lacks support from other students.	.06	.10	**.62**	.39

Note. Boldface indicates highest factor loadings.

■ **Table 16.4.**

Here spanners are included in the table (i.e., "Factor 1: External . . . ," "Factor 2: Student Effort," "Factor 3: Support . . ."). These are capitalized because they include factors.

Table X

Factor Loadings for Varimax Orthogonal Three-Factor Solution for the Items of the Student Failure Questionnaire

Item	Factor loading
Factor 1: External Reasons (Out of Student's Control)	
3. Education taught in a second language.	.89
14. Student has learning disability.	.86
9. Student has attention deficit disorder.	−.86
5. Student lacks adequate diet.	.79
11. Student appears to be of superior intellect.	.79
1. Student is excessively absent from school.	.77
Factor 2: Student Effort	
7. Student does not do homework.	.86
15. Student enjoys doing schoolwork.	.79
2. Student does extra work on his or her own.	.77
13. Student makes extra effort to participate in class.	.75
4. Student asks questions about material.	.69
Factor 3: Support From Others	
12. Student lacks parental support.	.90
6. Student lacks support from school.	.89
8. Student lacks support from teacher.	.75
10. Student lacks support from other students.	.62

Note. $N = 81$ and $\alpha = .76$ for entire measure.

Only the highest factor loading for each item is presented (i.e., a single factor loading for each item is shown rather than three factor loadings). In this example, factor loadings greater than .40 are considered high (the specific cutoff used varies according to the researcher).

Sometimes the researcher may include additional information in the table. Tables 16.5 through 16.7 provide different examples of how this can be achieved. In Table 16.5, item means for boys and girls within the sample are provided along with the factor loadings. Table 16.6 provides item means and standard deviations for the entire sample, rotated item factor loadings for all three factors, and item communalities. Table 16.7 reports the internal consistency of the items that form a factor and the factor loadings. (Internal consistency information regarding a set of items is useful to determine whether the items are to be considered as separate scales. This information can be presented within the text.)

Table 16.5.

Table X

Item Means for Respondents and Factor Loadings From Principal Component Factor Analysis With Varimax Rotation for the Student Failure Questionnaire

Item	M Boys	M Girls	Factor loading
Factor 1: External Reasons (Out of Student's Control)			
3. Education is taught in a second language.	5.23	5.39	.89
14. Student has learning disability.	6.44	6.70	.86
9. Student has attention deficit disorder.	5.29	4.55	−.86
5. Student lacks adequate diet.	4.81	5.13	.79
11. Student appears to be of superior intellect.	3.73	3.55	.79
1. Student is excessively absent from school.	4.52	3.80	.77
Factor 2: Student Effort			
7. Student does not do homework.	5.27	4.91	.86
15. Student enjoys doing schoolwork.	5.44	4.80	.79
2. Student does extra work on his or her own.	4.56	3.52	.77
13. Student makes extra effort to participate in class.	4.40	4.86	.75
4. Student asks questions about material.	5.01	4.65	.69
Factor 3: Support From Others			
12. Student lacks parental support.	5.64	5.10	.90
6. Student lacks support from school.	3.50	3.04	.89
8. Student lacks support from teacher.	3.41	3.51	.75
10. Student lacks support from other students.	2.70	3.16	.62

Note. N = 81. Item mean scores reflect the following response choices: 1 = strongly disagree, 2 = moderately disagree, 3 = slightly disagree, 4 = neither agree nor disagree, 5 = slightly agree, 6 = moderately agree, and 7 = strongly agree.

■ **Table 16.6.**

Table X

Means, Standard Deviations, Rotated Factor Loadings, and Communalities for Student Failure Questionnaire Items

Item	M	SD	Factor loadings			h^2
			1	2	3	
3	5.31	1.61	**.89**	.02	−.12	.80
14	6.57	2.45	**.86**	−.02	.02	.74
9	4.92	1.82	**−.86**	.11	.13	.76
5	4.97	2.17	**.79**	.04	−.14	.65
11	3.64	2.98	**.79**	.12	.03	.64
1	4.16	2.01	**.77**	−.16	.08	.62
7	5.09	1.39	.05	**.86**	−.08	.76
15	5.12	2.67	.01	**.79**	.10	.63
2	4.04	1.28	.03	**.77**	.09	.60
13	4.63	2.91	.10	**.75**	.32	.67
4	4.83	1.30	−.09	**.69**	.32	.59
12	5.37	2.56	.09	.09	**.90**	.82
6	3.27	1.74	.09	.11	**.89**	.81
8	3.46	1.43	.08	.06	**.75**	.58
10	2.93	1.38	.06	.10	**.62**	.39

Note. Boldface indicates highest factor loadings. Description of items found in Appendix A. Factor 1 = External Reasons (Out of Student's Control); Factor 2 = Student Effort; Factor 3 = Support From Others; h^2 = communality.

Here the researcher indicates in the note that item descriptions are presented in an appendix (not presented here). Alternatively, this could be mentioned in the body of the text.

■ **Table 16.7.**

Coefficient alphas calculated based on all of the
items that make up a factor are provided.

Table X

*Principal-Components Analysis With Varimax Rotation and
Coefficient Alphas for the Student Failure Questionnaire Items*

Item	Factor loadings
Factor 1: External Reasons (Out of Student's Control) ($\alpha = .70$)	
3	.89
14	.86
9	−.86
5	.79
11	.79
1	.77
Factor 2: Student Effort ($\alpha = .62$)	
7	.86
15	.79
2	.77
13	.75
4	.69
Factor 3: Support From Others ($\alpha = .54$)	
12	.90
6	.89
8	.75
10	.62

Note. Item descriptions can be found in Appendix A.

This is the "Play It Safe" table for an obliquely rotated factor analysis solution.

Here the researcher indicates in the note that item descriptions are presented in an appendix (not presented here).
Alternatively, this could be mentioned in the body of the text.

Example 16.3

In this example, the SFQ is factor analyzed using an oblique (oblimin) rotation (Table 16.8). Because an oblimin rotation is used, the researcher should present the correlations among the factors as illustrated in Table 16.8.

Table 16.8.

Table X

Summary of Factor Loadings for Oblimin Three-Factor Solution for the Student Failure Questionnaire

Item	Factor loading 1	2	3
1. Student is excessively absent from school.	.65	.85	.32
2. Student does extra work on his/her own.	.34	.90	.54
3. Education is taught in a second language.	.88	.41	.03
4. Student asks questions about material.	.34	.55	.91
5. Student lacks adequate diet.	.99	.34	.53
6. Student lacks support from school.	.45	−.02	.78
7. Student does not do homework.	.57	.85	.44
8. Student lacks support from teacher.	.24	.43	.91
9. Student has attention deficit disorder.	.89	.68	.32
10. Student lacks support from other students.	.43	.42	.72
11. Student appears to be of superior intellect.	.92	.31	.02
12. Student lacks parental support.	.42	.38	.68
13. Student makes extra effort to participate in class.	.34	.55	.24
14. Student has learning disability.	.45	.46	.34
15. Student enjoys doing schoolwork.	.34	.63	.03
Factor correlations			
Factor 1	—		
Factor 2	.39	—	
Factor 3	.26	.31	—

Table 16.9.

Only the highest factor loadings are presented in this table.

Table X

Student Failure Questionnaire Item Factor Loadings (× 100): Oblimin Rotation

Item	External Reasons	Student Effort	Support From Others
		Factor loadings	
1		.85	
2		.90	
3	.88		
4			.91
5	.99		
6			.78
7		.85	
8			.91
9	.89		
10			.72
11	.92		
12			.68
13		.55	
14	.45		
15		.63	
		Factor correlations	
Factor 1	—		
Factor 2	.39	—	
Factor 3	.26	.31	—

Note. Only the highest factor loadings are presented.

Example 16.4

In this example, the SFQ was given to two separate samples: (a) teachers and (b) parents. The factor structure of the questionnaire was determined for each set of raters. Tables 16.10 and 16.11 are examples of the presentation of the results.

▨ **Table 16.10.**

> This is the "Play It Safe" table for the results of two separate factor analyses.

Table X

Factor Scores and Communalities for Student Failure Questionnaire Items Based on the Varimax-Rotated Three-Factor Solution

| | Teachers ($n = 81$) | | | | Parents ($n = 213$) | | | |
| | Factor | | | | Factor | | | |
Item	1	2	3	h^2	1	2	3	h^2
3	**.89**	.02	−.12	.80	**.62**	−.21	.11	.45
14	**.86**	−.02	.02	.74	**.58**	.23	.32	.49
9	**−.86**	.11	.13	.76	**−.65**	.22	.27	.55
5	**.79**	.04	−.14	.65	**.70**	.38	.45	.84
11	**.79**	.12	.03	.64	**.62**	.10	−.29	.47
1	**.77**	−.16	.08	.62	**.59**	.17	.42	.55
7	.05	**.86**	−.08	.76	.32	**.71**	.10	.62
15	.01	**.79**	.10	.63	.43	**.69**	.21	.70
2	.03	**.77**	.09	.60	−.22	**.83**	.37	.87
13	.10	**.75**	.32	.67	.41	**.56**	.02	.49
4	−.09	**.69**	.32	.59	.29	**.73**	−.04	.61
12	.09	.09	**.90**	.82	.23	.12	**.52**	.34
6	.09	.11	**.89**	.81	.03	.23	**.69**	.53
8	.08	.06	**.75**	.58	.43	−.09	**.72**	.71
10	.06	.10	**.62**	.39	.16	.09	**.64**	.45

Note. Boldface indicates highest factor loadings. Descriptions of items are found in Appendix A.
h^2 = communality.

> Here the researcher indicates in the note that item descriptions are presented in an appendix (not presented here). Alternatively, this could be mentioned in the body of the text.

■ **Table 16.11.**

For each sample, items have been ordered according to their factor loadings.

Table X

Factor Scores for Student Failure Questionnaire Items for Teacher and Parent Raters

	Teachers		Parents	
Factor	Item	Factor loading	Item	Factor loading
1: External Reasons (Out of Student's Control)				
	3	.89	5	.70
	14	.86	9	−.65
	9	−.86	3	.62
	5	.79	11	.62
	11	.79	14	.58
	1	.77	1	.59
2: Student Effort				
	7	.86	2	.83
	15	.79	4	.73
	2	.77	7	.71
	13	.75	15	.69
	4	.69	13	.56
3: Support From Others				
	12	.90	8	.72
	6	.89	6	.69
	8	.75	10	.64
	10	.62	12	.52

Note. Item descriptions can be found in Appendix A.

Here the researcher indicates in the note that item descriptions are presented in an appendix (not presented here). Alternatively, this could be mentioned in the body of the text.

17

Multiple Regression

What Is It?

Multiple regression is a statistical procedure that assesses the relation between one criterion (dependent) variable and several predictor (independent) variables. There are several types of multiple regression analyses (e.g., stepwise, forward, backward), which differ in the way independent variables are entered into the regression equation.

What Tables Are Used?

There is considerable variability in the way results from multiple regression analyses are presented. That is, there are many versions of multiple regression tables in the literature. This variability is based on the purpose of the particular multiple regression table. Some researchers consider more aspects of multiple regression analyses than others and hence provide more comprehensive tables. Typically, when reporting the results of multiple regression analyses, two tables are used: (a) a table of means, standard deviations, and correlations for predictor and dependent variables (Table 17.1) and (b) a multiple regression analysis summary table (Tables 17.2–17.8).

"Play It Safe" Table

The "Play It Safe" table for a standard multiple regression analysis is Table 17.2. The "Play It Safe" table for a hierarchical multiple regression analysis is Table 17.4. Note also that to be "safe," a table of means, standard deviations, and intercorrelations should be included (Table 17.1).

Example 17.1

Researchers were examining the effect of parent characteristics on children's early phonemic awareness. Phonemic awareness tends to predict reading success, and the researchers wanted to determine parent characteristics related to that early awareness. Their study included 68 preschool children and their parents. The dependent (criterion) variable is phonemic awareness, as measured by a test the researchers called the Phonemic Awareness Measure (PAM). On the PAM, children are asked to choose, for example, which words start with the same sound. The independent (predictor) variables are the parent characteristics: education level, literacy, hours per week of own reading, hours per week of reading to the child, and socioeconomic status.

Variables for Example 17.1

Independent Variables

1. Parental education level
2. Parental literacy
3. Hours per week of parents' own reading
4. Hours per week of reading to child
5. Socioeconomic status

Dependent Variable

1. Phonemic awareness (Phonemic Awareness Measure score)

▨ **Table 17.1.**

There are many alternate formats for correlation tables and tables of means and standard deviations. See Chapters 7 and 3, respectively.

Table X

Means, Standard Deviations, and Intercorrelations for Children's Phonemic Awareness and Parent Characteristics Predictor Variables

Variable	M	SD	1	2	3	4	5
Phonemic Awareness Measure	43.56	9.34	.72***	.54***	.04	.46***	.58***
Predictor variable							
1. Parental education level	12.92	2.71	—	.67***	−.01	.38**	.79***
2. Parental literacy	10.21	2.19		—	−.05	.10	.62***
3. Parents' own reading	3.40	1.44			—	−.00	.06
4. Reading to child	2.06	0.51				—	.30*
5. Socioeconomic status	9.88	1.38					—

*$p < .05$. **$p < .01$. ***$p < .001$.

▨ **Table 17.2.**

This is the "Play It Safe" table for standard multiple regression analysis.

Table X

Regression Analysis Summary for Parent Variables Predicting Children's Phonemic Awareness

Variable	B	$SE\ B$	β	t	p
Parental education level	1.84	0.53	.53	3.51	.001
Parental literacy	0.75	0.50	.18	1.51	.136
Parents' own reading	0.37	0.54	.06	0.68	.497
Reading to child	4.60	1.68	.25	2.75	.008
Socioeconomic status	−0.20	0.93	−.03	−0.21	.831

Note. $R^2 = .58$ ($N = 68$, $p < .001$). ◄┈┈┈┈

Note that either the adjusted or unadjusted R^2 value may be presented.

┈┈► The R^2 value could be presented in the text rather than in a table note.

Table 17.3.

Table X

Regression Analysis Summary for Parent Variables Predicting Children's Phonemic Awareness

Variable	*B*	95% CI	β	*t*	*p*
Parental education level	1.84	[−9.06, 15.78]	.53	3.51	.001
Parental literacy	0.75	[0.79, 2.89]	.18	1.51	.136
Parents' own reading	0.37	[−0.71, 1.44]	.06	0.68	.497
Reading to child	4.60	[1.25, 7.95]	.25	2.75	.008
Socioeconomic status	−0.20	[−2.07, 1.67]	−.03	−0.21	.831

Note. $R^2 = .58$ ($N = 68$, $p < .001$). CI = confidence interval for *B*.

This table shows 95% confidence intervals for *B*.

Table 17.4.

This is the "Play It Safe" table for a hierarchical multiple regression analysis.

Table X

Hierarchical Regression Analysis Summary for Parent Variables Predicting Children's Phonemic Awareness (N = 68)

Step and predictor variable	*B*	*SE B*	β	R^2	ΔR^2
Step 1:				.52***	
Parental education level	2.50	0.29	.72***		
Step 2:				.53***	.01
Parental literacy	0.42	0.49	.10		
Step 3:				.53***	.00
Parents' own reading	0.34	0.56	.05		
Step 4:				.58***	.05**
Reading to child	4.57	1.66	.25**		
Step 5:				.58***	.00
Socioeconomic status	−0.20	0.93	−.03		

**$p < .01$. ***$p < .001$.

■ **Table 17.5.**

Table X

Prediction by Parent Predictor Variables of Children's Phonemic Awareness

Hierarchical step	Predictor variable	Total R^2	Incremental R^2
1	Parental education level	.52***	
2	Parental literacy	.53***	.01
3	Parents' own reading	.53***	.00
4	Reading to child	.58***	.05**
5	Socioeconomic status	.58***	.00

**p < .01. ***p < .001.

■ **Table 17.6.**

For this table, the analysis was changed (in terms of the variables included in each step) to illustrate inclusion of semipartial correlations.

Table X

Hierarchical Regression Analysis Predicting Phonemic Awareness With Parent Variables

Step and predictor variable	R^2	ΔR^2	sr	β
Step 1	.53***	.53***		
Parental education level			.48***	.66***
Parental literacy			.08	.10
Parents' own reading			.05	.05
Step 2	.55***	.02		
Reading to child			.23**	.25**
Socioeconomic status			−.02	−.03

Note. sr = semipartial correlation coefficient.
**p < .01. ***p < .001.

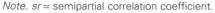

.53 in ΔR^2 column is redundant; it reflects that nothing was changed.

Table 17.7.

Table X

Summary of Hierarchical Multiple Regression Analysis With Children's Phonemic Awareness as Criterion

Step	Predictor variable	R^2	ΔR^2	ΔF	p
1	Parental education level	.52	.52	72.70	<.001
2	Parental literacy	.53	.01	0.72	.399
3	Parents' own reading	.53	.00	0.37	.546
4	Reading to child	.58	.05	7.61	.008
5	Socioeconomic status	.58	.00	0.05	.831

The .52 in ΔR^2 column demonstrates that this is the first entry and that no other variable has been entered.

Table 17.8.

This is the briefest version of a multiple regression table.

Table X

Hierarchical Multiple Regression Analysis Relating Parent Variables to Children's Phonemic Awareness

Step and predictor variable	β	ΔR^2
Parental education level	.72	.52***
Parental literacy	.10	.01
Parents' own reading	.05	.00
Reading to child	.25	.05**
Socioeconomic status	−.03	.00

** $p < .01$. *** $p < .001$.

Logistic Regression

18

What Is It?

Logistic regression is a variant of multiple regression; the procedure assesses the relation between one criterion (dependent) variable and several predictor (independent) variables. In logistic regression the criterion variable is categorical, and the predictor variables usually include both categorical and continuous variables. Logistic regression analysis allows the researcher to estimate the odds of an event (one level of the dependent variable) occurring on the basis of the values for the predictor variables.

What Tables Are Used?

There are three tables that are most commonly used to report the results of a logistic regression analysis: (a) a table of means and frequencies (Table 18.1), (b) a table of intercorrelations for predictor and criterion variables (Table 18.2), and (c) a logistic regression summary table (Table 18.3 or 18.4).

"Play It Safe" Table

The "Play It Safe" table for logistic regression is Table 18.3. Note that to be "safe," a table of means and frequencies (Table 18.1) and a table of intercorrelations (Table 18.2) should be included.

Example 18.1

A group of researchers conducted a study investigating factors related to childhood amnesia. *Childhood amnesia* refers to the lack of early childhood memories demonstrated by most adults. The researchers hypothesized that childhood amnesia should be less severe for adults who as young children spent a considerable amount of time talking and interacting with adults. The participants were 140 college students. Half of the college students could recall at least one event that happened before they were 3 years 6 months old, and the other half of the students had no memories of events that happened before they were 3 years 6 months old. Thus, participants in the first group of students had an early childhood memory, whereas participants in the second group of students did not. These are the two levels of the criterion (dependent) variable. The predictor (independent) variables are (a) whether the participant was an only child, (b) frequency of visits with grandparents, (c) IQ, (d) verbal fluency, and (e) working memory span.

Variables for Example 18.1

Independent Variables

1. Whether only child
2. Frequency of visits with grandparents
3. IQ
4. Verbal fluency
5. Working memory span

Dependent Variable

1. Early childhood memory (presence or absence of)

Table 18.1.

For other examples of tables of means and standard deviations, frequencies, and a priori contrasts, see Chapters 3, 2, and 6, respectively.

Table X

Mean Values or Frequencies for Predictor Variables as a Function of Early Childhood Memory

Variable	Memory before 3 years 6 months (n = 70)	No memory before 3 years 6 months (n = 70)	$\chi^2(1)$ or $t(138)$	p
Only child (%)	71	39	15.27	<.001
Grandparent visits	8.26	5.83	5.43	<.001
IQ	111.60	102.71	8.87	<.001
Verbal Fluency score	46.34	50.04	−3.29	.001
Working memory span	8.26	7.47	2.73	.007

Note. Chi-square test used for only child variable; *t* test used for all other variables.

Table 18.2.

Table X

Intercorrelations for Early Childhood Memory and Predictor Variables

Measure	1	2	3	4	5	6
1. Early childhood memory	—					
2. Only child	.33***	—				
3. Grandparent visits	.42***	.18*	—			
4. IQ	.60***	.06	.32***	—		
5. Verbal Fluency score	−.27**	−.14	.03	−.21*	—	
6. Working memory span	.23**	.14	.19*	.23**	−.08	—

Note. Early childhood memory coded as 1 = memory before age 3 years 6 months, 0 = no memory before age 3 years 6 months. Only child coded as 1 = only child, 0 = not only child.
*p < .05. **p < .01. ***p < .001.

■ **Table 18.3.**

This is the "Play It Safe" table for a logistic regression summary.

Table X

Summary of Logistic Regression Analysis Predicting Early Childhood Memory

Variable	B	SE	OR	95% CI	Wald statistic	p
Only child	−1.91	0.56	0.15	[0.05, 0.44]	11.57	.001
Grandparent visits	0.36	0.12	1.43	[1.14, 1.80]	9.54	.002
IQ	0.25	0.05	1.29	[1.17, 1.42]	26.88	<.001
Verbal Fluency score	−0.11	0.04	0.90	[0.83, 0.97]	6.83	.009
Working memory span	0.08	0.16	1.08	[0.79, 1.48]	0.25	.620

Note. CI = confidence interval for odds ratio (*OR*).

■ **Table 18.4.**

Table X

Logistic Regression Predicting Early Childhood Memory

Predictor	B	SE	p	OR	95% CI
Only child	−1.91	0.56	.001	0.15	[0.05, 0.44]
Grandparent visits	0.36	0.12	.002	1.43	[1.14, 1.80]
IQ	0.25	0.05	<.001	1.29	[1.17, 1.42]
Verbal Fluency score	−0.11	0.04	.009	0.90	[0.83, 0.97]
Working memory span	0.08	0.16	.620	1.08	[0.79, 1.48]

Note. CI = confidence interval for odds ratio (*OR*).

19

Confirmatory Factor Analysis

Confirmatory Factor Analysis: What Is It?

The purpose of confirmatory factor analysis is to test the goodness of fit of one or more hypothesized factor models of a measure. The models frequently consist of *observed variables* (those that are measured, often the participants' responses to particular items on an instrument) identified as *indicator variables, observed indicators, measured variables,* or *items* of a measure. These indicator variables are often grouped together to form one or more factors. These factors are the latent constructs or factors that have been hypothesized by the researcher and are thought to best describe the measurement instrument.

What Tables Are Used?

Two types of tables are used: (a) a summary table of the fit indices for each hypothesized model (see Tables 19.1 and 19.2) and (b) a table of the factor loadings of the best-fit model (see Table 19.3). If only one or two models are being tested, then the fit indices may be presented within the text rather than in a table.

Other information such as the means and standard deviations of the items of the measure, coefficient alphas for the scales, and interscale correlations also could be presented. See Chapter 3 of this volume for examples of tables of means and standard deviations and Chapter 7 for examples of correlation tables.

Example 19.1 presents the results of a confirmatory factor analysis for a single sample. Example 19.2 presents the results for two samples.

"Play It Safe" Table

There are two comprehensive tables: (a) the summary table listing the goodness-of-fit indices and (b) the table of the factor loadings (i.e., the standardized regression weights) for the best-fit model. If only one or two models are tested, the fit indices table is not needed; that information can be provided within the text. If more than two models have been tested, then a table of the fit indices should be presented. Tables 19.1, 19.3, 19.4, and 19.5 are "Play It Safe" tables.

Example 19.1

A new 15-item questionnaire to measure flexibility of thought was created. The questionnaire was given to 400 university students. The researchers wished to determine which of their three hypothesized factor structures best described this new measure. The three models are as follows:

Model 1—A single-factor model;

Model 2—A two-factor model consisting of (a) Ability to Adjust and (b) Openness to Ideas; and

Model 3—A three-factor model consisting of (a) Ability to Compromise, (b) Openness to Ideas, and (c) Ability to Adjust.

Variable for Example 19.1

1. The 15 items of the Flexibility of Thought Questionnaire

Table 19.1.

A summary table such as this is important, particularly if more than two models are being tested. If only one or two models are being tested, then the results can be given within the text.

Table X

Goodness-of-Fit Indices of Four Models (N = 400)

Model	df	χ^2	χ^2/df	AGFI	ECVI	RMSEA
Null	105	555.61***	5.29			
Single factor	106	526.24***	4.96	.76	.75	.061
Two factor	107	225.18***	2.10	.80	.79	.054
Three factor	108	88.42	0.82	.92	.90	.034

Note. AGFI = adjusted goodness-of-fit index; ECVI = expected cross-validation index; RMSEA = root mean square error of approximation.
***$p < .001$.

If the change in chi-square is to be presented, it can be included in a column immediately to the right of the column labeled χ^2/df.

Other indices of fit, such as the normed fit index, nonnormed fit index, or normed comparative fit index, could be included in additional columns. Usually, three or more indices are included.

This table, along with a table of factor loadings, is the "Play It Safe" table for confirmatory factor analysis.

Table 19.2.

In this table, the null model is not included.

Table X

Goodness-of-Fit Indicators of the Models for the Flexibility of Thought Questionnaire (N = 400)

Model	df	χ^2	χ^2/df	AGFI	ECVI	RMSEA
Single factor	106	526.24***	4.96	.76	.75	.061
Two factor	107	225.18***	2.10	.80	.79	.054
Three factor	108	88.42	0.82	.92	.90	.034

Note. AGFI = adjusted goodness-of-fit index; ECVI = expected cross-validation index; RMSEA = root mean square error of approximation.
***$p < .001$.

■ **Table 19.3.**

This table of factor loadings for the best-fit model, along with a table of fit indices, is the "Play It Safe" table for presenting the results of a confirmatory factor analysis.

For additional examples of these kinds of tables, see Chapter 16.

Table X

Standardized Solutions by Confirmatory Factor Analysis for the Three-Factor Model

	Factor		
Item	Ability to Compromise	Openness to Ideas	Ability to Adjust
10	.58		
5	.57		
14	.56		
9	.44		
3	.39		
12		.52	
2		.51	
8		.49	
1		.45	
6		.40	
13			.55
4			.50
15			.48
7			.43
11			.39

Example 19.2

The Flexibility of Thought Questionnaire was given to two samples: (a) a group of students from a university located in a large city in Canada and (b) a group of students from an equally large city in the United States. The researchers wished to determine how their three models (as identified in Example 19.1) fit each of these samples. Table 19.4 presents the goodness-of-fit indices, and Table 19.5 presents the factor loadings for the best-fit model.

Table 19.4.

This table, along with a table of factor loadings, is the "Play It Safe" table for presenting the results of confirmatory factor analysis for two samples.

Table X

Goodness-of-Fit Indicators of Models for the Flexibility of Thought Questionnaire for Two Samples (N = 400)

Model	df	χ^2	χ^2/df	AGFI	ECVI	RMSEA
Sample 1						
Single factor	106	526.24***	4.96	.76	.75	.061
Two factor	107	225.18***	2.10	.80	.79	.054
Three factor	108	88.42	0.82	.92	.90	.034
Sample 2						
Single factor	106	502.37***	4.74	.78	.76	.060
Two factor	107	217.98***	2.04	.81	.80	.051
Three factor	108	79.35	0.73	.91	.91	.032

Note. AGFI = adjusted goodness-of-fit index; ECVI = expected cross-validation index; RMSEA = root mean square error of approximation.
***$p < .001$.

Other indices of fit could be included in additional columns.

Table 19.5.

Table X

Factor Loadings of Flexibility of Thought Questionnaire Items for the Three-Factor Model for Two Samples

	Ability to Compromise		Openness to Ideas		Ability to Adjust	
Item	S1	S2	S1	S2	S1	S2
10	.58	.53				
5	.57	.56				
14	.56	.49				
9	.44	.50				
3	.39	.40				
12			.52	.47		
2			.51	.51		
8			.49	.56		
1			.45	.38		
6			.40	.46		
13					.55	.53
4					.50	.47
15					.48	.45
7					.43	.58
11					.39	.45

Note. S1 = Sample 1; S2 = Sample 2.

For additional examples of these kinds of tables, see Chapter 16.

This table, along with a table of fit indices, is the "Play It Safe" table for presenting the results of a confirmatory factor analysis for two samples.

20

Structural Equation Modeling

Model Testing: What Is It?

The purpose of structural equation modeling, as it relates to model testing, is to test the goodness of fit of one or more hypothesized models. These hypothesized models present hypothesized relations between constructs (latent variables) and their observed variables (those that are measured) that serve as indicators of those constructs.

What Tables Are Used?

In showing results of model testing, it is important to provide the reader with a visual representation of the model or models being tested. Thus, figures frequently are used.

Two tables are commonly presented: (a) a table of intercorrelations among the variables included in the analysis and the means and standard deviations of those variables (see Tables 20.1 and 20.2) and (b) a table of the fit statistics (Table 20.3). In addition to these two types of tables, figures of the models often are included.

If only one model is being tested, a figure of the hypothesized model is often presented within the introduction section (see Figures 20.1–20.6). In addition, a figure of the structural model obtained from the results of the analysis is presented (see Figures 20.7–20.9). If more than one model is being tested, then either all of the models are presented before (in the introduction) and after (in the Results section) the analyses, or a figure of one hypothesized model and written descriptions of alternate models are presented within the introduction with only the best fitting structural model presented within the Results section after the analyses.

If only one or two models are presented, then a table of fit indices is not required. This information can be incorporated within the text of the Results section.

"Play It Safe" Table

Figures 20.1 and 20.7 and Tables 20.1 and 20.3 are required to be comprehensive.

Example 20.1

Researchers wished to determine whether emotional expressive behavior (EEB), operationalized as crying, laughing, and yelling, is influenced by an individual's acceptance of EEB (attitudes and beliefs), which in turn is influenced by primary EEB influences (significant other's, father's, and mother's EEB) and personality (reactive and impulsive traits). They tested three models. Each of the three models is presented in a figure in this chapter (Figures 20.1–20.3). Models 2 and 3 are variants of Model 1 and do not need to be presented in the introduction, but they should be described within the text.

Variables for Example 20.1

Latent Variables

1. Primary emotional expressive behavior (EEB) influences
2. Personality
3. Personal acceptance of EEB indicator variables
4. EEB

Indicator Variables for EEB Influences

1. Significant other's EEB
2. Father's EEB
3. Mother's EEB

Indicator Variables for Personality

1. Reactive personality trait
2. Impulsive personality trait

Indicator Variables for Personal Acceptance of EEB

1. Attitudes regarding EEB
2. Beliefs regarding EEB

Indicator Variables for EEB

1. Crying
2. Laughing
3. Yelling

Figure 20.1.

According to APA Style, figures are presented after tables in a manuscript. Each figure begins on a separate page and is described in a figure caption. Both figure and figure caption are included on the same page, with the caption below the figure.

This is Model 1. This is a "Play It Safe" figure.

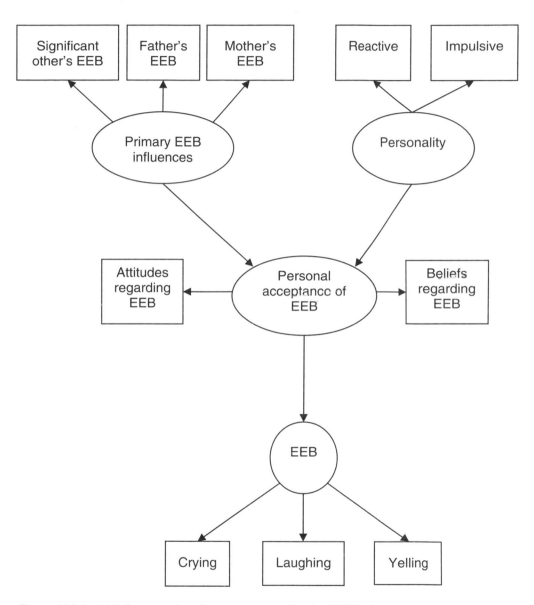

Figure X. Model 1 for emotional expressive behavior (EEB). Latent constructs are shown in ellipses, and observed variables are shown in rectangles.

Figure 20.2.

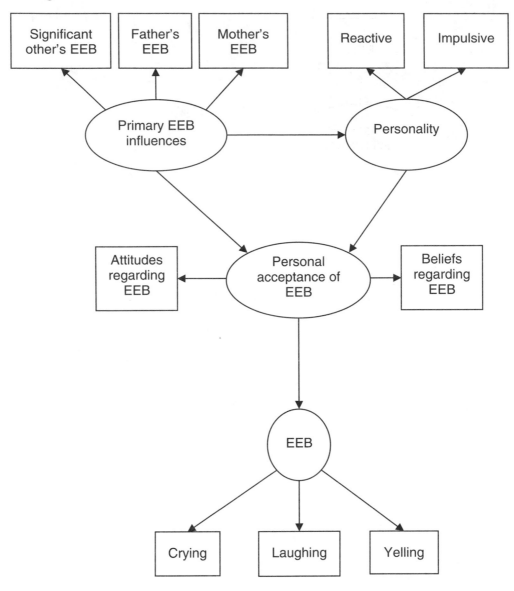

Figure X. Model 2 for emotional expressive behavior (EEB). Latent constructs are shown in ellipses, and observed variables are shown in rectangles.

Arrows may cross, but try to avoid this. Figures should be as simple as possible to facilitate comprehension.

Figure 20.3.

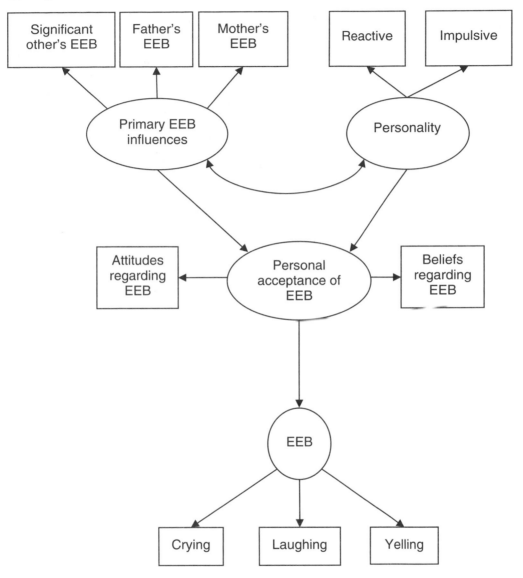

Figure X. Model 3 for emotional expressive behavior (EEB). Latent constructs are shown in ellipses, and observed variables are shown in rectangles.

If more than one figure is presented, the style should be similar, if not identical.

■ **Figure 20.4.**

Figures 20.4, 20.5, and 20.6 are
variants of Model 1.

In this figure, the researcher
wishes to focus attention solely
on the latent variables.

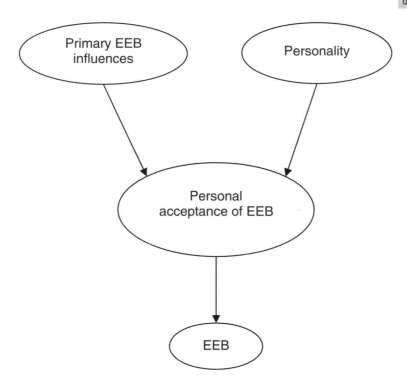

Figure X. Emotional expressive behavior (EEB), Model 1.

Figure 20.5.

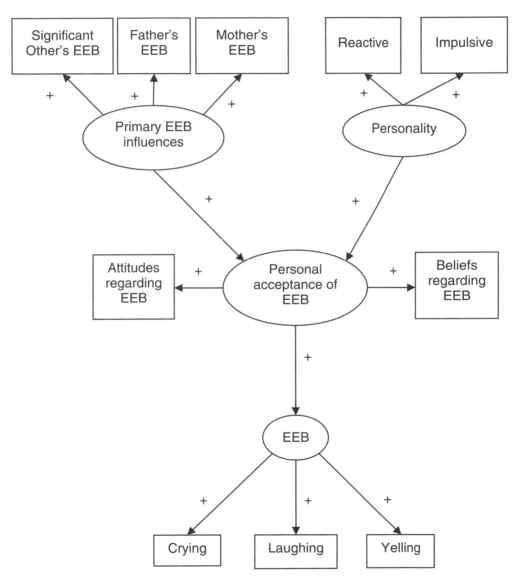

Indicating whether the relations are positive or negative is informative if they are not all positive or all negative. (In this model, all of the relations are positive, so it is not necessary to label the arrows as such.)

Figure X. Model 1 for emotional expressive behavior (EEB). Measurement error terms are not shown. Latent constructs are shown in ellipses, and observed variables are shown in rectangles.

■ **Figure 20.6.**

Using a symbol for the indicator variable rather than the name of the variable can save space if the diagram is complex and contains many latent or indicator variables, or both. The same symbols can then be used in the correlation table.

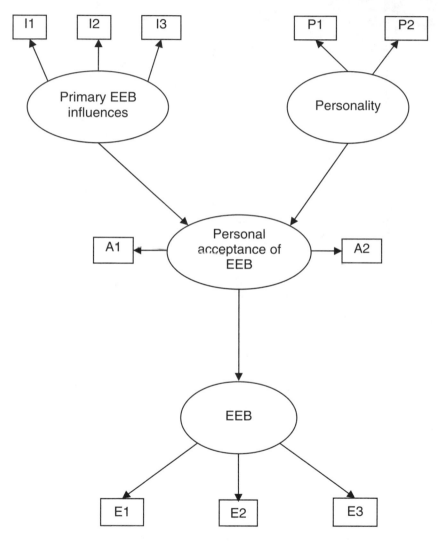

Figure X. Model 1 for emotional expressive behavior (EEB). Latent constructs are shown in ellipses, and observed variables are shown in rectangles. I1 = significant other's EEB; I2 = father's EEB; I3 = mother's EEB; P1 = reactive; P2 = impulsive; A1 = attitudes; A2 = beliefs; E1 = crying; E2 = laughing; E3 = yelling.

Table 20.1.

This table, along with a table of fit indices and illustrations of the models, is the "Play It Safe" table for model testing. This table provides headings to clearly identify the variables.

A table of intercorrelations among the variables along with the variable means and standard deviations is usually presented in the Results section.

See Chapters 7 and 3, respectively, for other examples of format for correlation tables and tables of means and standard deviations.

Table X

Descriptive Statistics and Zero-Order Correlations for Indicator Variables

Variable	1	2	3	4	5	6	7	8	9	10
				Primary EEB influences						
1. Mother	—									
2. Father	.26	—								
3. Significant other	.05	.09	—							
				Personality						
4. Reactive	.67	.56	.44	—						
5. Impulsive	.34	.24	.11	.21	—					

(continues)

Table 20.1. (continued)

Table X

Descriptive Statistics and Zero-Order Correlations for Indicator Variables

Variable	1	2	3	4	5	6	7	8	9	10
					Acceptance of EEB					
6. Attitudes	.65	.66	.23	.54	.21	—				
7. Beliefs	.06	.08	.15	.21	.12	.18	—			
				EEB						
8. Crying	.76	.32	.21	.17	.19	.14	.28	—		
9. Laughing	.51	.45	.41	.43	.21	.43	.65	.34	—	
10. Yelling	.49	.37	.35	.34	.12	.48	.51	.29	.10	—
M	45.67	35.61	54.31	12.45	13.56	32.65	23.55	65.49	26.34	43.25
SD	8.45	7.86	9.54	1.34	2.54	6.48	4.55	10.12	2.36	8.31

Note. Correlations greater than .19 are significant at *p* < .05. EEB = emotional expressive behavior.

■ **Table 20.2.**

Labels in this table correspond with those in Figure 20.6.

Table X

Zero-Order Correlations, Means, and Standard Deviations for Study Variables

Variable	I1	I2	I3	P1	P2	A1	A2	E1	E2	E3
I1	—									
I2	.09	—								
I3	.05	.26	—							
P1	.44	.56	.67	—						
P2	.11	.24	.34	.21	—					
A1	.23	.66	.65	.54	.21	—				
A2	.15	.08	.06	.21	.12	.18	—			
E1	.21	.32	.76	.17	.19	.14	.28	—		
E2	.41	.45	.51	.43	.21	.43	.65	.34	—	
E3	.35	.37	.49	.34	.12	.48	.51	.29	.10	—
M	54.31	35.61	45.67	12.45	13.56	32.65	23.55	65.49	26.34	43.25
SD	9.54	7.86	8.45	1.34	2.54	6.48	4.55	10.12	2.36	8.31

Note. Correlations greater than .19 are significant at $p < .05$. I1 = significant other's emotional expressive behavior (EEB); I2 = father's EEB; I3 = mother's EEB; P1 = reactive; P2 = impulsive; A1 = attitudes; A2 = beliefs; E1 = crying; E2 = laughing; E3 = yelling.

■ **Table 20.3.**

This table, along with diagrams of the models and a table of correlations and means and standard deviations of the variables, is the "Play It Safe" table for model testing.

Table X

Fit Statistics for Alternative Models

Model	df	$\chi^2(N = 300)$	GFI	CFI	RMSR	IFI
1	44	187.65*	.81	.79	.78	.041
2	43	184.24*	.80	.80	.79	.044
3	43	79.23	.91	.91	.90	.021

Note. GFI = goodness-of-fit index; CFI = comparative fit index; RMSR = root-mean-square residual; IFI = incremental fit index.
*$p < .01$.

Other indices of fit could be presented as well, such as the normed fit index. It is best to present three or more indices.

It is important to include a summary table of the fit indices, particularly if more than two models are being tested.

A diagram of the best fit model could be included (in this instance, Model 3; see Figures 20.7–20.9). Alternatively, three diagrams (one for each of the three models, including their standardized path coefficients) could be included.

◼ **Figure 20.7.**

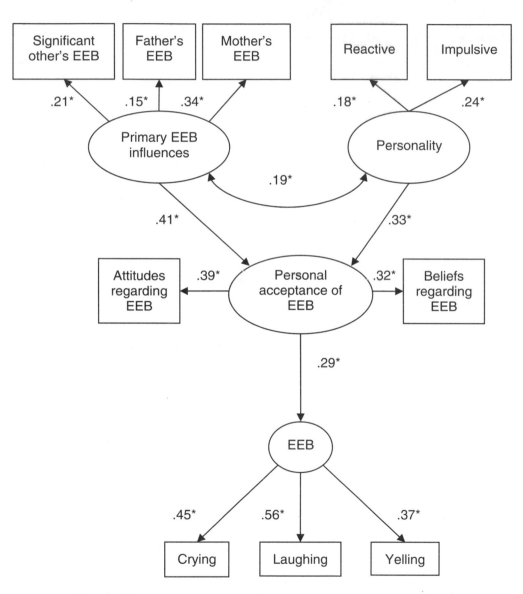

This is the "Play It Safe" figure for the best-fit model.

Figure X. Standardized coefficients for Model 3. Latent constructs are shown in ellipses, and observed variables are shown in rectangles. EEB = emotional expressive behavior. *p* < .05.

Figure 20.8.

Inclusion of fit indices within the figure is useful, particularly if only one model is being tested and a fit statistics table will not be generated.

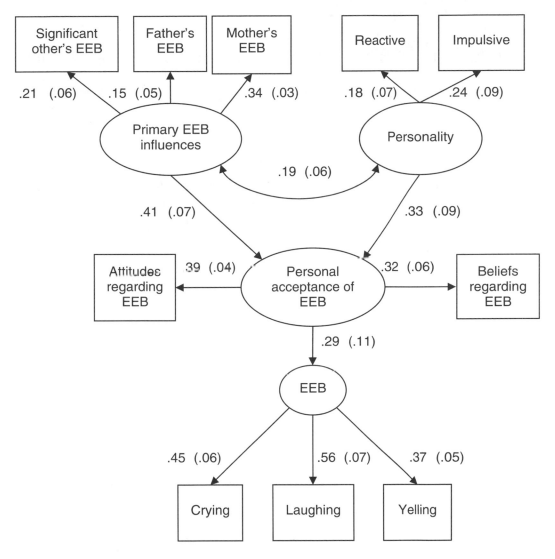

Figure X. Standardized coefficients for Model 3 and their standard errors (in parentheses). Latent constructs are shown in ellipses, and observed variables are shown in rectangles. All coefficients are significant at $p < .05$. $\chi^2(43, N = 300) = 79.23$, $p > .05$; goodness-of-fit index = .91; comparative fit index = .91; root mean square residual = .90; incremental fit index = .021. EEB = emotional expressive behavior.

Figure 20.9.

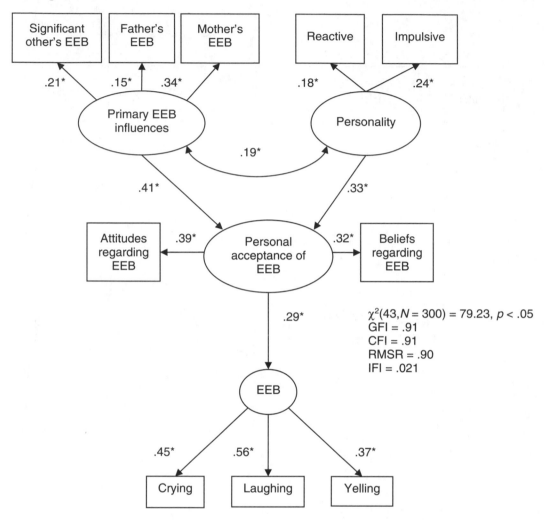

Figure X. Standardized coefficients for Model 3. Latent constructs are shown in ellipses, and observed variables are shown in rectangles. GFI = goodness of fit; CFI = comparative fit index; RMSR = root-mean-square residual; IFI = incremental fit index. EEB = emotional expressive behavior. *p < .05.

21

Meta-Analysis

What Is It?

Meta-analysis is a statistical technique that is used to combine the results of several independent studies. Meta-analysis typically is used to describe the population distribution of an effect size, a correlation, or a reliability coefficient.

What Tables Are Used?

Most researchers use two tables to present the results of a meta-analysis: (a) a table that summarizes the characteristics of the studies included in the meta-analysis and (b) a table of the results of the meta-analysis. Sometimes, however, researchers present only the first type of table. This is usually the case if the researcher is interested only in the overall effect size, reliability, or correlation of the results in the meta-analysis and is not interested in how the effect size might differ depending on the studies' various moderating characteristics. These researchers will present one table containing the characteristics of the studies and usually will provide the overall effect size, reliability, or correlation in the text (see Example 21.1 and Tables 21.1 and 21.2). Researchers who present both types of tables are interested in whether the effect size, reliability, or correlation differs in different categories of studies (i.e., how the effect size is affected by various moderating variables) and use the second type of table to present those findings (see Example 21.2 and Tables 21.3–21.6).

"Play It Safe" Table

If a researcher is interested only in the overall effect size, reliability, or correlation computed in the meta-analysis and has not made any hypotheses about how these might vary depending on different characteristics of the previous studies, then Table 21.2 is the "safe" choice. If, however, the researcher has hypotheses about moderating variables or different categories of studies, then both Tables 21.2 and 21.5 constitute the "Play It Safe" choice.

Example 21.1

In this example, the researchers were interested in attitude change. Specifically, they were interested in determining how attitudes are affected by endorsement from an attractive versus an unattractive speaker. There have been many previous studies on this aspect of attitude change, and these researchers wanted to perform a meta-analysis on the results of those studies to establish whether there is a significant overall effect for attitude change depending on endorser attractiveness. After searching different research databases (e.g., PsycINFO), the researchers chose 15 studies that satisfied their criteria for inclusion in the meta-analysis. (Note that all of the studies used in this example are fictitious.)

For this example, the dependent variable is attitude change, which has been assessed in many different ways in the studies examined by the researchers. The independent variable is the attractiveness of the speaker or endorser.

Variables for Example 21.1

Independent Variable

1. Endorser attractiveness (attractive vs. unattractive)

Dependent Variable

1. Attitude change

■ **Table 21.1.**

The summary table is not required by APA Style. This type of table could be included as supplemental material. Alternatively, studies included in the meta-analysis can be identified with asterisks in the reference list (if the meta-analysis includes fewer than 50 studies), or a list of studies can be included as supplemental material (if the meta-analysis includes more than 50 studies).

Table X

Summary of Studies Included in Meta-Analysis on Attitude Change and Endorser Attractiveness

Study	n	Cohen's d	SD
Bigness and Holmes (2002)	56	0.20	0.21
Brank and Jones (1991)	145	1.32	0.98
Calvert and Snell (1987, Experiment 2)	69	0.24	0.22
Cruikshank et al. (1996)	77	1.85	1.12
Crewski (2008)	88	1.22	0.88
Dunkley, Rogers, and Graham (1992)	102	0.17	0.09
Goodchild and Harding (2007a)	100	1.70	1.45
Goodchild and Harding (2007b, Experiment 3)	56	1.44	1.41
Goodchild and Kacinik (2003)	50	2.08	1.55
Harrison (2004, Experiment 1)	156	0.37	0.23
Harrison et al. (2005)	98	0.46	0.44
Jenkins and Harrison (1990)	28	0.24	0.21
Melville and Harrison (1999)	58	1.28	0.88
Smith and Crane (2001)	122	0.91	0.55
Smith and Smith (1998, Experiment 2)	78	0.94	0.66

In this sample table, *d* is used to express effect size, but there are other measures that might be presented instead (e.g., effect size, *Zr*), or the researcher could present reliability or correlation coefficients rather than (or in addition to) effect size.

▩ Table 21.2.

Because the outcome measures varied among the studies included in this meta-analysis, it could be useful to list these different measures in the summary table, as illustrated in this table.

This is the "Play It Safe" version of a summary table for a meta-analysis.

Table X

Summary of Studies Included in Meta-Analysis on Attitude Change and Endorser Attractiveness

Study	Indicator of attitude change	n	Cohen's d	95% CI LL	95% CI UL
Bigness and Holmes (2002)	Opinion on affirmative action	56	0.20	0.10	0.30
Brank and Jones (1991)	Opinion on standardized tests	145	1.32	1.12	1.42
Calvert and Snell (1987, Experiment 2)	Opinion on abortion	69	0.24	0.11	0.37
Cruikshank et al. (1996)	Opinion on spanking children	77	1.85	0.11	0.37
Crewski (2008)	Opinion on universal health care	88	1.22	1.42	2.28
Dunkley, Rogers, and Graham (1992)	Opinion on euthanasia	102	0.17	0.99	1.45
Goodchild and Harding (2007a)	Opinion on capital punishment	100	1.70	1.45	1.95
Goodchild and Harding (2007b, Experiment 3)	Opinion on capital punishment	56	1.44	1.23	1.63
Goodchild and Kacinik (2003)	Opinion on capital punishment	50	2.08	1.78	2.38
Harrison (2004, Experiment 1)	Opinion on Title IX	156	0.37	0.27	0.47
Harrison et al. (2005)	Opinion on affirmative action	98	0.46	0.31	0.61
Jenkins and Harrison (1990)	Opinion on abortion	28	0.24	0.15	0.33
Melville and Harrison (1999)	Opinion on abortion	58	1.28	1.12	1.44
Smith and Crane (2001)	Opinion on abortion	122	0.91	0.71	1.11
Smith and Smith (1998, Experiment 2)	Opinion on abortion	78	0.94	0.69	1.19

Note. CI = confidence interval; *LL* = lower limit; *UL* = upper limit.

Example 21.2

In this study, the researchers were interested in attitude change and endorser attractiveness, but they also were interested in how this relationship would be moderated by several variables. They hypothesized that effect size might be moderated by (a) the sex of the endorser relative to the sex of the participant (endorsers who are a different sex than the participants will be most effective), (b) the age of the endorser relative to the age of the participant (endorsers who are older than the participants will be more effective), and (c) the mode of presentation for the endorsement (endorsements in person will be more effective than endorsements on video recording).

For this example, the researchers would include a summary table of the individual studies included in the meta-analysis (such as Table 21.1 or 21.2), but they would also include a table of the meta-analysis results (such as Tables 21.3–21.7).

■ Table 21.3.

Table X

Summary Statistics for Total Sample in Meta-Analysis and for Three Moderating Variables

Category	n	k	Cohen's d	p	SD
Total sample	1,273	15	0.92	<.001	0.23
Sex of target versus endorser					
Other	240	4	0.96	<.001	0.21
Same	362	8	0.25	.068	0.15
Age of target versus endorser					
Younger	244	3	0.83	.002	0.15
Same	204	4	0.36	.016	0.18
Older	455	6	1.44	<.001	0.21
Presentation mode					
Video recorded	256	5	0.40	.009	0.21
In person	828	10	1.06	<.001	0.21

Note. k = number of samples.

Standard deviations for effect size are not always included in these tables and are sometimes expressed in different ways (e.g., σ_δ).

Table 21.4.

Table X

Summary Statistics for Effects of Sex, Age, and Presentation Mode in Meta-Analysis

Category	k	Cohen's d	p_d	95% CI LL	95% CI UL	Q	p_Q
Sex of target versus endorser							
Other	4	0.96	<.001	0.80	1.12	5.36	.207
Same	8	0.25	.068	0.12	0.38	10.40	.008
Age of target versus endorser							
Younger	3	0.83	.002	0.60	1.06	3.20	.440
Same	4	0.36	.016	0.22	0.40	9.64	.019
Older	6	1.44	<.001	1.22	1.66	2.11	.574
Presentation mode							
Video recorded	5	0.40	.009	0.30	0.50	4.91	.256
In person	10	1.06	<.001	0.90	1.22	8.80	.036

Note. CI = confidence interval; k = number of samples; p_d = significance of d; *LL* = lower limit; *UL* = upper limit; Q = Cochran's measure of homogeneity; p_Q = *significance of Q.*

■ **Table 21.5.**

This is the "Play It Safe" table for presenting meta-analysis results.

Table X

Mean Effect Sizes for Moderators

Category	n	k	Cohen's d	p_d	95% CI LL	95% CI UL	r	Q	p_Q
Sex of target versus endorser									
Other	240	4	0.96	<.001	0.80	1.12	.49	5.36	.207
Same	362	8	0.25	.068	0.12	0.38	.13	10.40	.008
Age of target versus endorser									
Younger	244	3	0.03	.002	0.60	1.06	.35	3.20	.440
Same	204	4	0.36	.016	0.22	0.40	.18	9.64	.019
Older	455	6	1.44	<.001	1.22	1.66	.71	2.11	.574
Presentation mode									
Video recorded	256	5	0.40	.009	0.30	0.50	.21	4.91	.256
In person	828	10	1.06	<.001	0.90	1.22	.51	8.80	.036

Note. CI = confidence interval; k = number of samples; p_d = significance of d; LL = lower limit; UL = upper limit; Q = Cochran's measure of homogeneity; p_Q = *significance of Q.*

If a different type of meta-analysis is conducted, then some different statistics might be included: the mean sample size-weighted effect size (M_{wt}) and standard deviation (SD_{wt}), the unweighted mean (M_{unwt}) and standard deviation (SD_{unwt}), and (for reliabilities) the mean and standard deviation for the square root of the effect size estimates, both weighted (M_{sqwt}, SD_{sqwt}) and unweighted (M_{squnwt}, SD_{squnwt}).

■ **Table 21.6.**

Table X

Summary of Effect Sizes for Moderating Variables of Sex, Age, and Presentation Mode

Category	k	Cohen's d	p_d	r	Q	p_Q
Sex of target versus endorser						
Other	4	0.96	<.001	.49	5.36	.207
Same	8	0.25	.068	.13	10.40	.008
Age of target versus endorser						
Younger	3	0.83	.002	.35	3.20	.440
Same	4	0.36	.016	.18	9.64	.019
Older	6	1.44	<.001	.71	2.11	.574
Presentation mode						
Video recorded	5	0.40	.009	.21	4.91	.256
In person	10	1.06	<.001	.51	8.80	.036

Note. k = number of samples; p_d = significance of d; Q = Cochran's measure of homogeneity; p_Q = *significance of Q.*

This is the most common table for illustrating moderating effects in meta-analysis because these are the four statistics that are most commonly presented. However, the columns are not always in the order presented here.

Table 21.7.

This is the briefest table for presenting meta-analysis results.

Table X

Summary Statistics for Three Moderating Variables in Meta-Analysis

			95% CI	
Category	Cohen's *d*	*p*	*LL*	*UL*
Sex of target versus endorser				
Other	0.96	<.001	0.80	1.12
Same	0.25	.068	0.12	0.38
Age of target versus endorser				
Younger	0.83	.002	0.60	1.06
Same	0.36	.016	0.22	0.40
Older	1.44	<.001	1.22	1.66
Presentation mode				
Video recorded	0.40	.009	0.30	0.50
In person	1.06	<.001	0.90	1.22

Note. CI = confidence interval; *p* = significance of *d; LL* – lower limit; *UL* = upper limit.

Word Tables

22

What Is It?

A *word table* provides descriptive or qualitative information. The types of information included in word tables include definitions of variables, descriptions of referenced studies, and order of presentation of training phases.

What Tables Are Used?

Usually, one table is used for each type of descriptive information presented. Word tables are used sparingly; a word table should be included only if a thorough description is required and if a description in the text would be too confusing or too cumbersome.

"Play It Safe" Table

Because of the varied nature of word tables, there is no specific "Play It Safe" word table.

Example 22.1

Two researchers wished to determine which variables from a set were most strongly related to their Emotional Well-Being Scale in a population of individuals 65 years of age and older. There were many variables, so the researchers determined that they would be presented most effectively in a word table. Tables 22.1–22.3 show several different ways of presenting the information.

Variables for Example 22.1

Independent Variables

1. Spouse support
2. Child support
3. Friend support
4. Pet companionship
5. Routine living
6. Social activities
7. Physical activities
8. Mental activities
9. Income
10. Resource availability
11. Financial support

Dependent Variable

1. Emotional well-being

Table 22.1.

Table X

Definitions of Variables and Sample Items

Variable	Definition	Sample item (positively keyed)
Spouse support	Spouse is emotionally supportive.	My partner is always willing to listen to me.
Child support	Child(ren) is/are emotionally supportive.	I can always count on my child(ren) to assist me with my problems.
Friend support	Friend(s) is/are emotionally supportive.	My friends call me on a weekly basis to keep in touch.
Pet companionship	Pet provides comfort.	I enjoy talking to my pet.
Routine living	Day-to-day living is predictable or repetitive.	It is important that I keep my routine every day.
Social activities	Individual participates in social activities.	I belong to a club where I can meet people.
Physical activities	Individual participates in physical activities.	I enjoy playing organized sports.
Mental activities	Individual participates in mental activities such as learning, reading, or problem solving.	I try to read every day.
Income	Level of income available to individual.	My average income per month is ____.
Resource availability	Number and quality of resources available to individual.	I have easy access to a grocery store.
Financial support	Financial support from family and friends.	I can rely on family or friends when I have money problems.

▇ Table 22.2.

If definitions for the various variables are provided in the text of a manuscript, then it is not necessary to present them in the table. Here, sample items are presented. Items that are reverse scored are identified with an R.

Table X

Example Questionnaire Items for the 11 Variables

Variable	Example questionnaire items
Spouse support	My partner is always willing to listen to me.
	My partner is not always there for me. (R)
Child support	I can always count on my child(ren) to assist me with my problems.
	I gain strength from being with my child(ren).
Friend support	My friends call me on a weekly basis to keep in touch.
	I don't feel comfortable enough to talk with my friends about personal issues. (R)
Pet companionship	I enjoy talking to my pet.
	I never feel alone because I have a pet.
Routine living	I try to do something new every day. (R)
	It is important that I keep my routine every day.
Social activities	I belong to a club where I can meet people.
	I prefer not to do volunteer work. (R)
Physical activities	I enjoy playing organized sports.
	I try to get some exercise every day.
Mental activities	I try to read every day.
	I avoid writing as much as possible. (R)
Income	My average income per month is ____.
Resource availability	I have easy access to a grocery store.
	I live close to my doctor's office or a hospital.
Financial support	I can rely on family or friends when I have money problems.
	I have a lot of savings in the bank.

Note. (R) denotes items that are reverse scored.

Table 22.3.

> Sometimes, additional information regarding the variables or questionnaires is provided in the table.

Table X

Definitions of Variables and Sample Items

Variable	Sample item	No. of items	Cronbach's α
Spouse support	My partner is always willing to listen to me.	10	.91
Child support	I can always count on my child(ren) to assist me with my problems.	7	.87
Friend support	My friends call me on a weekly basis to keep in touch.	8	.77
Pet companionship	I enjoy talking to my pet.	4	.67
Routine living	It is important that I keep my routine every day.	5	.64
Social activities	I belong to a club where I can meet people.	8	.71
Physical activities	I enjoy playing organized sports.	6	.59
Mental activities	I try to read every day.	6	.60
Income	My average income per month is ____.	1	
Resource availability	I have easy access to a grocery store.	10	.76
Financial support	I can rely on family or friends when I have money problems.	6	.78

Example 22.2

Researchers had access to data provided by a group of female participants in their 20s and in their 40s, and they had the same participants (now in their 70s) complete the Emotional Well-Being Measure. Because the scales used in the various periods of time differed, the researchers wished to present much of the descriptive information in a table. See Table 22.4.

Variables for Example 22.2

Variables for 20-Year-Olds

1. Sadness scale
2. Overall intelligence score

Variables for 40-Year-Olds

1. Number of years worked
2. Marital status
3. Number of children
4. Happiness scale
5. Depression scale

Variables for 70-Year-Olds

1. Emotional Well-Being Scale
2. Number of years worked
3. Marital status
4. Number of children
5. Number of grandchildren
6. Number of deaths experienced in the family

Table 22.4.

Table X

Descriptive Data for the Samples on the Various Measures Taken at Three Different Points of Time

Sample	Descriptive data
20-year-olds	Sadness scale: $M = 22.3$, $SD = 5.6$
	Overall intelligence score: $M = 104.2$, $SD = 7.5$
40-year-olds	Number of years worked: $M = 8.4$, $SD = 4.5$
	Marital status: 53.1% married, 12.1% common-law relationship, 34.8% single
	Number of children: $M = 2.4$, $SD = 0.3$
	Happiness scale: $M = 15.3$, $SD = 3.5$
	Depression scale: $M = 10.1$, $SD = 7.5$
70-year-olds	Emotional Well-Being Scale: $M = 25.6$, $SD = 6.1$
	Number of years worked: $M = 24.7$, $SD = 8.9$
	Marital status: 25.9% married, 6.1% common-law relationship, 68.0% single
	Number of children: $M = 2.6$, $SD = 0.3$
	Number of grandchildren: $M = 2.9$, $SD = 1.2$
	Number of deaths experienced in the family: $M = 12.5$, $SD = 4.3$

Index

About the Authors

Adelheid A. M. Nicol, PhD, received her doctorate in industrial/organizational psychology from the University of Western Ontario, London, Ontario, Canada, in 1999. She is an associate professor in the Military Psychology and Leadership Department at the Royal Military College of Canada, Kingston, Ontario, Canada. Her current research interests are in the area of prejudice and industrial/organizational psychology. She teaches courses in English and in French in cross-cultural psychology, industrial psychology, organizational psychology, personality, research methods, and social psychology.

Penny M. Pexman, PhD, earned her doctorate in psychology from the University of Western Ontario, London, Ontario, Canada, in 1998. She is now a professor in the Department of Psychology at the University of Calgary, Calgary, Alberta, Canada. In her research, she examines several aspects of language processing in adults and in children, including word recognition processes and figurative language understanding. She is an award-winning teacher and graduate supervisor.